Risk levels in coastal and river engineering

A guidance framework for design

I. D. Mockett
J. D. Simm

Published by Thomas Telford Publishing, Thomas Telford Ltd, 1 Heron Quay, London, E14 4JD

URL: http://www.thomastelford.com

Distributors for Thomas Telford books are
USA: ASCE Press, 1801 Alexander Bell Drive, Reston, VA 20191-4400, USA
Japan: Maruzen Co. Ltd, Book Department, 3–10 Nihonbashi 2-chome, Chuo-ku, Tokyo 103
Australia: DA Books and Journals, 648 Whitehorse Road, Mitcham 3132, Victoria

First published 2002

A catalogue record for this book is available from the British Library

ISBN: 0 7277 3164 7

Typeset by Alex Lazarou, Surbiton, Surrey
Printed and bound in Great Britain by MPG Books, Bodmin, Cornwall

Preface

In practice, modern coastal and fluvial engineering design involves a combination of consensus building with stakeholders, which can be viewed as a process of setting appropriate risk levels. If, through more detailed consideration of risks and their acceptability, the engineering community could make a 5% saving in the whole-life costs of coastal structures, then the monetary benefit would be significant. For example, it has been estimated that a whole-life cost saving of £50–100 million annually could be achieved in the UK alone and approximately £200 million annually where UK designers are working overseas.

The development of concepts and methods to achieve these savings has been the aim of this report funded by the UK government through both the Department for Trade and Industry (DTI) and the Health and Safety Executive (HSE), with the focus on multi-attribute design and setting acceptable risk levels. This research has reinforced the concept that efficient design not only requires good technical analysis but also needs to consider the social aspects of design. The stakeholders in a particular project all have a personal perspective on the project objectives and it is the designer's challenge to manage these multiple concerns and aspirations efficiently. If the efficiency of decision making can be improved then it is quite possible that this 5% or larger saving can be achieved.

This book is aimed primarily at the designer in order to develop a robust philosophy for undertaking and managing the design of coastal and fluvial structures. However, it has also been written to help other decision makers to broaden their understanding of the design process so that efficient decisions regarding coastal and fluvial engineering can be achieved. The book has been split into two parts: Part A is the main guidance setting out the framework for setting design risk levels, while

Part B provides the case study analysis used to develop and demonstrate the main principles in Part A.

PART A — ESTABLISHING A RISK FRAMEWORK FOR COASTAL AND FLUVIAL ENGINEERING

The two issues that have become prominent through this research are as follows.

A conceptual model combining multiple attributes and risk

Although the primary focus of coastal engineers is understandably on scientific and technical issues, there are actually a wide range of risks or concerns that need to be considered, including, economics, environmental impact, political influence, etc. A risk concerns the likelihood of an event that an individual does not wish to occur and, therefore, any risk will be unwelcome. However, to make rational decisions, risks or concerns can be banded into three distinct levels of 'broadly acceptable', 'tolerable' and 'unacceptable' (HSE, 1999).

In practice, the majority of risks fall in the tolerable region, in which case they should be managed so that they are *as low as reasonably practicable* (ALARP). Currently, the tolerability of risks is assessed intuitively for softer issues, such as tourism, and formally for more technical issues where, for example, indicative standards of protection are used (MAFF, 1999). However, there is still a need to develop a risk framework that integrates both soft and technical risks into a single integrated process.

Through back analysis of completed coastal engineering case studies, the concept of the 'Acceptable Risk Bubble' emerged as an attempt to visualise the process of managing, in an integrated way, risk tolerability against a range of concerns (see Figure P.1). The concept of the bubble is relatively simple: the centre of the bubble is defined as that which is totally acceptable, which — given that a risk will always be unwelcome — is a point that is never likely to be achieved, and the outer surface is defined as that which is unacceptable. Each axis from the centre represents a particular risk or concern controlled by a stakeholder. The stakeholder can define the units of the axis either qualitatively or quantitatively. For example, the economic axis may use the Benefit Cost Ratio as a measure, while the tourism axis may have more descriptive labels.

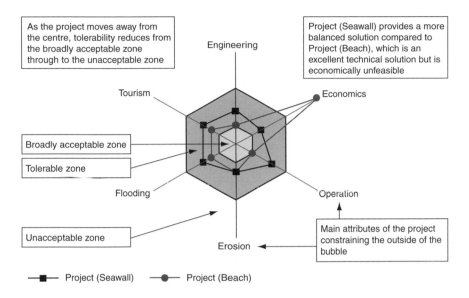

As the project moves away from the centre, tolerability reduces from the broadly acceptable zone through to the unacceptable zone

Engineering

Project (Seawall) provides a more balanced solution compared to Project (Beach), which is an excellent technical solution but is economically unfeasible

Tourism

Economics

Broadly acceptable zone

Tolerable zone

Flooding

Operation

Unacceptable zone

Erosion

Main attributes of the project constraining the outside of the bubble

▪ Project (Seawall)   ● Project (Beach)

Figure P.1.   An example plot of the Acceptable Risk Bubble

This approach allows the designer to set out the main issues involved in the project and aids communication between all stakeholders. Through the case study analysis, it was observed that most stakeholders would categorise risks as being either broadly acceptable or unacceptable at the start of the given project. It is not until the stakeholder considers other concerns or receives more information that some of the risks move from unacceptable to tolerable, demonstrating the social process of design through negotiation and compromise.

Development of a philosophy to understand and define a common risk level

The current approach to the management of risk in design and the setting of risk levels in coastal and fluvial engineering varies depending on the perspective of the individual, organisation and the dominant design culture in the project. Various approaches are used by different design disciplines, ranging from heavily codified procedures to reliance on best practice. The three main design disciplines involved in the design of coastal and fluvial structures are:

- *structures* — heavily codified, the use of partial safety factors
- *geotechnics* — often empirical design codes; limited use of partial safety factors
- *hydraulics* — limited codes, largely dependent on guidance notes.

The independent development of these approaches has resulted in minimal formal understanding of how an individual risk level in one design culture communicates with another. Poor communication of risk levels between design disciplines may lead to inappropriate levels of design risk being considered for different elements of a scheme, resulting in an inefficient design or, in the worst case, an unsafe design.

The initial aim of the research was to seek and define a common risk level in which coastal and fluvial structures should aspire, by reviewing the range of risk levels currently adopted by designers through the analysis of case studies. Understanding the risk levels for each structural element of the overall system and then how these risk levels interact as a system is not a trivial problem. The clear message of the research is that a prescribed common risk level for all coastal and fluvial structures would fail to work in practice because of the multi-attribute nature discussed earlier.

The future development of design has two strands. First, there is a real need to move towards the use of similar terminology (i.e. move towards lifetime probabilities rather than return period events or factors of safety). Second, the designer needs to consider a more systems based approach to design in order to develop an acceptable risk level for the overall system rather than the same risk level for each element. The conclusion of the research is that the designer can use indicative targets of lifetime probabilities as a starting point but, in essence, the risk levels need to be optimised appropriately for each individual project. As the design becomes more systems orientated then a larger flexibility exists in managing the design through its own lifetime by moving beyond arbitrary lifetime to open-ended streams of cost and benefit.

This book provides guidance on how the designer can move towards a more systems orientated design approach.

PART B — DETAILED CASE STUDY ANALYSIS

The development of Part A was supported by the analysis of a wide range of port, coastal and fluvial engineering case studies. These provide information on the decision-making process and the influence that stakeholders have on the overall design process. Issues that are discussed in the case study analysis include:

- application of the Acceptable Risk Bubble
- distribution of risk levels and their interaction within a structural system
- selection of one piece of coastline to focus on against another
- temporal variations in risk
- common risk based approach to design of a coastal slope
- dealing with different levels of tolerability
- spatial variations in risk
- selection of design life and design events
- comparison between analysing risk using different loading conditions
- finding an optimum risk level for bridge scour design
- acceptance of higher risk levels by the client.

It should be noted that HR Wallingford is responsible for any additional risk-related analysis discussed in Part B (typically in the Acceptable Risk Issues section of each case study). However, in most cases it was also appropriate to describe the essence of the design work originally undertaken by others; the reader should treat these descriptions as a summary of the key points rather than as a complete design report.

Ian Mockett graduated in 1996 from Heriot-Watt University with an MEng in Civil Engineering with European Studies and joined HR Wallingford in early 1997 as a Graduate Engineer within its Coastal Department. In 2001, he qualified as a Chartered Civil Engineer and was then seconded to the HR Wallingford Regional Office in Kuala Lumpar, Malaysia, to take on his current role as Technical Manager for Coastal and Maritime Engineering in SE Asia.

Jonathan Simm (j.simm@hrwallingford.co.uk) is Technical Director for Engineering at HR Wallingford. He has 25 years' experience of both project and research work in coastal engineering, and spent 14 years with engineering consultants before joining HR Wallingford in 1992. He is well known as the author and editor of a number of engineering manuals for coastal and river engineering.

Editor's note

Please note that Railtrack was in operation at the time of writing. The company has been replaced by Network Rail at the time of going to press.

Acknowledgements

This book has been produced as a result of an HR Wallingford Research Project carried out under the Department of Trade and Industry (DTI) 'Partners in Technology' scheme. (Note, the Contract was started under the Department of the Environment, Transport and the Regions (DETR) and transferred to the DTI in June 2001.) The objective of the project was to review and set out a framework in which a common risk based approach to hydraulic design could be adopted.

This book is published on behalf of the DTI and the Health and Safety Executive (HSE). The views and information presented in the guide are those of HR Wallingford and while they reflect the views of the advisory committee, they are not necessarily those of the funding organisations. Ian Mockett and Jonathan Simm of HR Wallingford managed the project and edited the book.

PROJECT FUNDERS

Financial support for the project was provided by:

- the DTI
- the HSE.

IN-KIND CONTRIBUTIONS

In-kind contributions, in terms of data on historic projects and staff time committed to the project were provided by the following organisations:

- Association of British Insurers
- Bristol University
- Environment Agency
- Health and Safety Executive
- High-Point Rendel
- Posford Duvivier
- Railtrack
- Technical University of Delft.

STEERING GROUP

An advisory committee to guide the research comprised the following members:

Noel Beech	Posford Duvivier
Jackie Bennett	Association of British Insurers
Malcolm Birkenshaw	HSE
Pieter van Gelder	Delft University of Technology
Jim Hall	Bristol University
Alan Hardie	Railtrack
John Mason	Alan Baxter and Associates (DTI Project Officer)
Ian Meadowcroft	Environment Agency
Ian Mockett	HR Wallingford (Project Manager)
Jonathan Simm	HR Wallingford (Chairman)
Hugo Wood	High-Point Rendel

Assistance was also received from the following people and organisations:

William Allsop	HR Wallingford
Ian Cooke	Posford Duvivier
Ian Cruickshank	HR Wallingford
Steve Fort	High-Point Rendel
Peter Haigh	Railtrack
Peter Martin	High-Point Rendel
Neal Masters	HR Wallingford
Chris Mounsey	Formerly of the Association of British Insurers
Koo-Yong Park	Hyundai (formerly of Oxford University)

| Maria Reis | Posford Duvivier |
| Folkert Schoustra | Delft University of Technology |

HR Wallingford are grateful for the support given to the project by the funders, the members of the advisory committee and all those organisations and individuals involved in the success of the project. We would also like to thank Elma Mann and Helen Stevenson for their patient and careful formatting of the manuscript, Bev Reader for her efforts with the correspondence and Carol Chedzey for her efforts in printing the original reports.

HR Wallingford is an independent specialist research, consultancy, software and training organisation that has been serving the water and civil engineering industries worldwide for over 50 years in more than 60 countries. We aim to provide appropriate solutions for engineers and managers working in:

- water resources
- irrigation
- groundwater
- urban drainage
- rivers
- tidal waters
- ports and harbour
- coastal waters
- offshore.

Address: Howbery Park, Wallingford, Oxon, OX10 8BA, UK
Internet: http://www.hrwallingford.co.uk

Please not that the authors of this book are employed by HR Wallingford. The work herein was carried out under a Contract jointly funded by HR Wallingford and the Secretary of State for Trade and Industry placed on 22 July 1999. Any views expressed are not necessarily those of the Secretary of State for Trade and Industry.

Glossary

Benefit Cost Ratio (BCR)	The ratio of the monetary benefit of undertaking the project (including the deduction of any residual losses) over the whole-life cost of undertaking the project divided by the capital expenditure to undertake the project.
Consequence	The outcome (or result) of an event, for example economic, social or environmental impact. May be expressed quantitatively (e.g. monetary value), by category (e.g. high, medium, low) or descriptively.
Decision/event/fault tree	Ways of describing a system and the linkages between different parts of the system. Useful for identifying causes, tracing possible sequences of events and investigating the effects of decisions.
Deterministic method	Method in which precise, single values are used for all variables and input values, giving a single value as the output.
Hazard	A situation with the potential to result in harm. A hazard does not necessarily lead to harm.

Coastal and fluvial design	The planning and detailed design of a coastal and fluvial system from need identification through to decommissioning.
Coastal and fluvial structure	A structure that is designed to withstand the forces operating in the coastal and fluvial environment.
Internal rate of return (IRR)	The discount rate that at which the net present value is equal to zero.
Monte Carlo modelling	A numerical technique for assessing the probability of different outcomes from two or more variables.
Net present value (NPV)	The difference between funds spent and revenue generated discounted over a period of time.
Probabilistic method	Method in which the variability of input values and the sensitivity of the results are taken into account to give results in the form of a range of probabilities for different outcomes.
Probability	The relative proportion or frequency of events leading to that outcome, out of all possible events.
Qualitative methods	Approaches which use descriptive rather than numerical values for assessment and decision making.
Quantitative methods	Approaches which use numerical values (or ranges of values) for assessment and decision making.
Residual risk	The risk that remains after risk management and mitigation. May include, for example, risk due to very severe (above the design event) storms or risks from unforeseen hazards.

Return period	The average length of time separating extreme events (e.g. a flood) of a similar magnitude.
Risk	The product of probability and consequence.
Risk assessment	Consideration of hazards inherent in a project and the risks associated with them.
Risk averse	The decision maker adopts the option with a more certain return even if the potential benefit is less.
Risk management	The activity of mitigating and monitoring risks, which occurs predominately after the project appraisal stage.
Risk neutral	The decision maker makes no distinction between probability and consequence when selecting an option.
Risk owner	An individual or organisation who is responsible for a specific risk.
Risk prone	The decision maker adopts the option that has the largest benefit even though it is less likely to occur (i.e. takes more of a gamble).
Risk register	An auditable record of the project risks, their consequences and significance, and proposed mitigation and management measures.
Risk workshop	A workshop where all parties involved in the project come together and discuss collectively the risks or concerns involved.
Sensitivity testing	Method in which the impact on the output of an analysis is assessed by systematically changing the input values.

Stakeholder   An individual or an organisation who has
an interest in a project.

Tolerability   The measure that an individual or
organisation will endure or tolerate a
specific risk.

Contents

Illustrations

Figures

Boxes

Introduction

1. Introduction

The following sections outline the aims, scope, readership, structure and use of the book.

1.1. THE NEED FOR A CLEARER UNDERSTANDING OF RISK LEVELS THAT ARE BEING ACCEPTED

Within the UK, coastal and fluvial structures have attracted adverse media attention through the occurrence of recent flooding, erosion and landslides. To combat potentially more frequent failures of coastal and fluvial structures, a more active approach to design and the setting of risk levels has been discussed within the engineering community for a number of years. If, through more detailed consideration of risks and their acceptability, the engineering community could make a 5% saving in the whole-life costs of coastal and fluvial engineering projects (including port engineering), then the monetary benefit would be between £50–100 million annually in the UK alone. The likely benefit to projects spread globally where UK designers are involved could be in the order of £250 million annually. The key question, therefore, is how can the designer actually make these savings. This book enables designers and clients to consider a philosophy to improve the efficiency of their designs, such that these savings may be achieved.

The current approach to the management of risk in design and the setting of risk levels in civil engineering varies depending on the perspective of the individual, organisation and the dominant design culture. Various approaches are used by different engineering disciplines, ranging from heavily codified procedures to the extensive use of best practice. The three main design disciplines generically involved in the design of coastal and fluvial structures, together with their typical design approach, are:

- *structures* — heavily codified, the use of partial safety factors
- *geotechnics* — often empirical design codes, limited use of partial safety factors
- *hydraulics* — limited codes, largely dependent on guidance notes.

The development of these different approaches has resulted in very little formal understanding of how an individual safety level in one design culture communicates with other cultures. Poor communication of risk levels may lead to inappropriate levels of design risk being considered for different elements of a scheme, resulting in an inefficient or unsafe design.

Technical analysis is an important element of the design process, but the coastal and fluvial engineer also needs to consider the social aspects of design as well, such as the influence of stakeholders. In some cases, the relevant stakeholders are not contacted in the early stages of the design process, resulting in late changes to meet the needs of a particular stakeholder, and hence the potential for inefficient design.

The aim of this book is to provide practical guidance on the process of setting acceptable risk levels within a specific project. However, this book is unable to set out definitive guidance on setting specific risk levels, since each project was individual and should be considered as thus.

1.2. OBJECTIVES OF THE BOOK

The principal objectives of this book are as follows:

- to provide clear guidance to clients, project funders, designers and other interested parties on the process of setting acceptable risk levels
- to illustrate the importance of involving all stakeholders in the setting of acceptable risk levels and throughout the design process
- to highlight the key areas that the designer must consider throughout the design process, including design life, design events and understanding the nature of specific risks

- to set out a layered approach to risk assessment that first considers the overall project and how it evolves as the project develops, then focuses on the actual risk levels accepted for a particular engineering option.

1.3. READERSHIP

This book is aimed at a wide readership. It is designed to serve and inform the needs of clients, project funders (e.g. the Department of the Environment, Food and Rural Affairs (DEFRA), the Environment Agency, local authorities, port authorities and private developers), consulting engineers and others interested in coastal and fluvial engineering (e.g. insurers). It is intended that the book be suitable for both the specialist and non-specialist. The book is not structure-specific so that the concepts and practices detailed within this book are applicable to a wide range of structures.

Clients and project funders will find useful information included on how to:

- define a project so that the key design issues can be readily identified and then tracked throughout the whole project cycle
- learn how to take responsibility for setting risk levels specific to the particular needs of the project.

Designers will find useful information included on how to:

- understand the importance of the various stages of the project cycle and the influence of stakeholders
- identify the specific risks to the project and how to manage them
- define the design life and use appropriate design events
- understand the importance of selecting the appropriate design tools and techniques.

Other interested bodies will find information on how they can bring influence during the stages of the project cycle.

Most of the issues discussed and current practice and policy described within this book relate to practice within the UK. However, a large amount of the information is also likely to be applicable to coastal and fluvial engineering on an international scale.

1.4. STRUCTURE AND USE OF THE BOOK

The book is split into the following two parts.

1.4.1. Part A

Part A sets out the framework in which risk levels should be defined and accepted. There are three main sections, which represent the three different levels of understanding that the designer needs to consider.

- *Defining a risk framework for integrating multi-attribute problems* (Chapter 2). Chapter 2 introduces the concept of risk/concern and how it can be categorised into tolerability levels varying from broadly acceptable, through tolerable, to unacceptable. The majority of coastal and fluvial engineering projects have a number of stakeholders who have an influence on the project's acceptability. The aim of the chapter is to set out the concept of framework in terms of an *Acceptable Risk Bubble* to assist in the efficient development of the project.
- *Management of the risk framework through the project process* (Chapter 3). Chapter 3 introduces the project cycle and discusses ways of incorporating the Acceptable Risk Bubble into that process. It also discusses the broad range of issues that need to be considered and highlights some techniques that designers can use during the design process.
- *Application of the risk framework on a specific engineering solution* (Chapter 4). Chapter 4 focuses on the engineering design of a specific engineering solution by summarising the key points that should be considered when setting acceptable risk levels for a specific engineering solution. It looks to the framework developed in Chapter 2 and managed in Chapter 3 as building blocks in setting acceptable risk levels. The chapter focuses on current acceptance of risk and promotes a framework in which the client, designer and others can work towards the delivery of an efficient, safe design that meets the requirements detailed by the stakeholders.

In essence, Part A provides a philosophy that builds on current good practice and looks towards improving the design process further in the future. Each chapter provides a preamble on the current thinking and then sets out a practical process that designers can apply to their own projects. References to case studies (Part B) are used to aid the reader's understanding of the issues that need to be considered. At the end of each chapter, there is a one-page summary of the key points to be considered during project development, or *aide mémoire*.

1.4.2. Part B

Part B contains the detailed case study analysis undertaken during the Department of Trade and Industry (DTI) research project to aid in the development of the framework described in Part A. This part provides more detailed information on certain risk issues that may be of specific interest to the reader.

The objectives of the case study analysis were to:

- gain an understanding of the influence of stakeholders in setting acceptable risk levels
- gain an understanding of the decision-making process during the design of coastal and fluvial structures
- demonstrate key principles that require consideration when setting risk levels in design.

Data for each case study were obtained through interviews with the consultants and clients involved with the individual projects. Further information was then collected using key references for each case study, such as feasibility studies and design reports. Individual references and acknowledgements are provided with each case study. Generally, the case study analysis is split into three main sections, as follows.

- *Influence of stakeholders on the final design.* This analysis considered how stakeholders in the project had influenced the final design and how, in some instances, they were prepared to compromise their position slightly through negotiation to reach a consensus on an acceptable design.
- *Decisions regarding setting acceptable risk levels during the design process at each stage.* This analysis considered where decisions on acceptable risk levels were made during the design process and how these decisions at different stages of the process influenced the final accepted risk level.

Table 1.1.   Summary of case study analysis

Case study	Description
1. Clacton coastal defences	A flood protection scheme with beach nourishment and the construction of new fishtailed groynes. The case study focuses on how the physical boundaries of a problem are defined and how risks should be assessed in terms of their change with time. The concept of the *Acceptable Risk Bubble* has been applied to this case study
2. Castle Cove coast protection	Protection of a coastal slope with improved drainage of the slope and the construction of a rock revetment to protect the toe. The case study focuses on the interaction between the hydraulic and geotechnical design disciplines. The concept of the *Acceptable Risk Bubble* has been applied to this case study as well
3. Dawlish seawall	Improvements to an existing vertical seawall that protects a main railway line. The case study focuses on the decision made by the client to address one risk, breaching, and to tolerate another, wave-induced overtopping. The concept of the *Acceptable Risk Bubble* has been applied to this case study
4. Grenada sea defences	Improvements to the coastal defences that protect the coastal main road through a mixture of rock revetments and cliff protection. The case study focuses on the spatial nature of risk by investigating a back analysis design check to ensure no sections of the design fell below the factor of safety used for the critical design cross-section
5. Tekirdag Port, Turkey	A new caisson breakwater to protect an existing port. The case study focuses on the setting of design return periods
6. West Bay coastal defence and harbour scheme	Refurbishment of an existing harbour. As part of the design process, innovative risk analysis was undertaken to gain a better understanding of the potential overtopping risks
7. Design of scour protection for bridges	The case study focuses on the issue of defining a set standard of protection against scour. It is argued that the designer should consider the consequence of failure in order to define the appropriate standard of protection
8. Acceptance of higher risk levels by the client	The case study focuses on the reconstruction of a fluvial tidal embankment that required the client to accept higher risk levels in order to provide a more appropriate solution

- *Acceptable risk issues*. This analysis sought to identify the challenges faced in setting risk levels for particular case studies so that they might have generic application to other projects.

The case studies cover a wide range of structures from river embankments, bridge abutments to port structures and coastal defence structures. A short summary of all the case studies presented is given in Table 1.1.

Part A

Establishing a risk framework for coastal and fluvial engineering

Defining a risk framework for integrating multi-attribute problems

2

2. Defining a risk framework for integrating multi-attribute problems

2.1. WHAT IS RISK?

Risk is generally understood to include elements of likelihood and consequence (Meadowcroft et al., 1997) and for convenience and simplicity is expressed as the product of these two elements, such that:

Risk = Probability × Consequence (or Benefit, then the risk would be seen as an opportunity)

In some situations, the predicted risk level for one option will be the same as another but the consequence may be higher and the probability lower or vice versa. For example, an option with a probability of 0·5 and potential earnings of £100 is just as risky as an option with a probability of 0·1 and potential earnings of £500. If the decision criterion is to select the least risky option then both are valid as the product of probability and consequence are both the same. However, if decision makers are more averse to risk then they will tend to select the first option, as it is more likely that they will receive something back in return, even if the sum is smaller.

So when considering risk, it is important to understand both probability and consequence, but decision making also involves a personal choice, which will largely depend on the particular situation and how averse the individual is to the given risk.

2.2   HOW DO WE CONSIDER RISKS?

In life, there are a number of risks or concerns that people choose to ignore (as being broadly acceptable), and others that they are not prepared to entertain (unacceptable). But there are also many risks or concerns that people are prepared to accept by weighing up the benefits of taking risks and the precautions that have to be taken to mitigate their undesirable effects. This promotes the concept of differing levels of tolerability between broadly acceptable and unacceptable risks.

The Health and Safety Executive (HSE, 1999) states that a risk concerns the likelihood of an event that an individual does not wish to occur and, therefore, any risk will be unwelcome. However, when making rational decisions, risks or concerns can be banded into three distinct levels which represent clearly the way individuals operate naturally. The three risk levels are summarised as being either one of the following.

- *Broadly acceptable* — level of residual risk regarded as insignificant and further effort to reduce risk not likely to be required, as resources to reduce risks are grossly disproportionate to the risk reduction achieved.
- *Tolerable or as low as reasonably practicable (ALARP)* — control measures must be introduced to drive residual risk towards the broadly acceptable region (e.g. maintenance and monitoring). If the residual risk remains above a broadly acceptable level and society desires the benefit of the activity, residual risk is only tolerable if further risk reduction is impractical or requires action that is grossly disproportionate in time, trouble and effort to lower the risk.

   Tolerable should not be confused with acceptable. It refers instead to a willingness to live with a risk so as to secure certain benefits and in the confidence that the risk is one that is worth taking and that it is properly controlled.
- *Unacceptable* — risk cannot be justified save in extraordinary circumstances.

The framework is illustrated in Figure 2.1. The triangle represents decreasing level of risk on moving from the top to the bottom. The HSE focuses primarily on individual risk and societal concerns (e.g. personal injury or fatalities), but this principle of acceptable risk

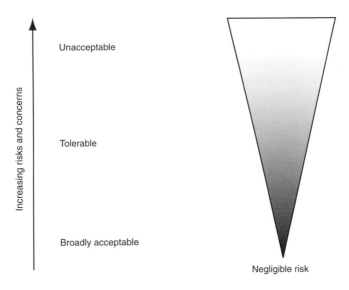

Figure 2.1. Criteria for the tolerability of risk (HSE, 1999)

regions and tolerability can be applied to all risks involved within coastal and fluvial engineering.

Taking the concept of tolerability a step further, decision-makers also need to understand how their own perception of risk may change between each of these three levels so that the efforts of the designer can be directed efficiently. This profile can be described as a value curve which indicates the rate of change in tolerability as a defined value is increased. The practical benefit of plotting a value curve for a particular risk or concern is that the designer can start to understand the sensitivity of the risk to change (i.e. small change in the design may make the project more or less acceptable to that particular attribute). Box 2.1 is an example of a potential value curve.

The concept of value curves moves us towards the mathematical realms of value and utility theory. Further information on value and utility theory is available from a number of sources but, as a first step, the interested reader is directed to French (1988).

Box 2.1.   An example of a value curve

The figure below shows the value curve of salary for an individual. What this curve demonstrates is how the perception of a value changes on money progressing up the curve. If the individual has a salary that is perceived to be unacceptable and if this curve is a static representation of how the individual's salary is perceived, then the step to a tolerable salary is a small step compared to the step from tolerable to broadly acceptable. Once the individual lies within the broadly acceptable region, then it is likely that the individual will either strive for a much larger salary increase or his efforts will be redirected to another value.

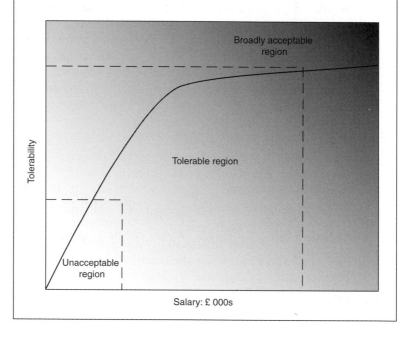

2.3.   THE TYPES OF GENERIC RISK INVOLVED IN HYDRAULIC DESIGN

Within the design of coastal and fluvial structures, there are a number of generic risks that may occur during the lifetime of the structure, the main categories being defined as:

- engineering
- financial
- economic
- insurance
- construction
- operation
- environment and heritage
- health and safety
- political
- societal.

These main factors and their implications for coastal and fluvial engineering are discussed further, reflecting the current views of the engineering community. As each project will have risks specific to its own situation, it is impossible to list all of the risks involved with coastal and fluvial engineering. The commentary below is presented as a starting point to highlight the key areas that the client and designer should consider as part of the design process.

2.3.1. Engineering

Any engineering scheme should be practical and certain designs may not be acceptable on engineering grounds as being, for example, too difficult to construct, inappropriate for the ground conditions, designed from materials unsuitable for the particular environment or based on inadequate provision for maintenance of the structure. Chapter 4 discusses the complexities of the engineering category in more detail.

2.3.2. Financial

Financial risk relates to whether money will be available when and where it is required. Generating funding for a project rarely involves the use of personal funds but rather securing funds through financial institutions or grant aid. The amount of funding made available will be based on the economics of the project (i.e. the benefit) and will involve convincing banks, government, local people or other financial institutions that the economics are sufficiently positive to allow funding. This topic is closely related to the issue of economic risk, which is discussed in the next section.

2.3.3. Economic

Economics can be defined as the study of production, distribution and consumption of wealth in human society. It is traditionally one of the key factors in defining acceptable risk levels because the economic indicators used are generally monetary based, which can be quickly integrated and compared with financial risk (i.e. funding). Acceptability is based on ensuring a given minimum level of benefits through the lifetime of the structure.

For example, a standard measure of economic performance is the Benefit Cost Ratio (BCR). In the UK, DEFRA provides grant aid to schemes for flood defence and coast protection work, with a BCR greater or equal to 1 with an appraisal period of 50 years. In practical terms, DEFRA will expect a much higher BCR because of the large number of schemes to prioritise when distributing their limited funds; the required value of the BCR and any other requirements are set out in their current *Grant Memorandum*.

2.3.4. Insurance

In terms of risk to society, the insurance industry is interested in making sure value for money is achieved through the construction of schemes that provide sufficient protection to insured assets but at an acceptable risk level that minimises overall exposure to risk. For example, the insurance industry would rather see a number of schemes put in place at a higher risk level than one scheme at a very low risk level. In this respect, the insurance industry has perhaps a larger influence on national policy rather than on the development of a scheme. However, this influence may change if, for example, individual insurance companies decide to withdraw flood cover from some areas that are frequently flooded.

From a different perspective, insurance companies also provide professional indemnity (PI) insurance for designers, and are keen therefore to ensure that designs do not fail because this would lead to heavy outlay. There may also be further payouts due to damage inflicted on the protected assets. During the research, a number of consultants highlighted the fact that some designs were strengthened significantly beyond the minimum requirements set out by design standards where they felt their PI insurance may be at risk and, therefore, their track record as designers.

2.3.5. Construction

Risks in construction are linked largely to issues regarding time, construction cost, safety and quality. More specific information can be found in two separate guidance manuals on specific risks involved with coastal and fluvial construction written for the DETR and edited by HR Wallingford. The two manuals are:

- *Construction risk in coastal engineering* (Simm and Cruickshank, 1998)
- *Construction risk in river and estuary engineering* (Morris and Simm, 2000).

2.3.6. Operation

The risks involved in the management of a structure once constructed should be considered. Issues include using the structure and allowing access for inspection and for maintenance with the ability to repair and replace elements if required. These risks are linked closely with health and safety risks

2.3.7. Environment and heritage

There is a growing recognition on the importance and value of the natural environment. Legislation at both national and international level has resulted in the need to consider the impacts of a scheme on the environment to ensure that any important recognised environmental designations are not harmed.

In the UK, as well as recognised habitats, there are also conservation and heritage site designations. These may include Areas of Outstanding Natural Beauty (AONBs), heritage coastlines, conservation areas, listed buildings, archaeology, etc.

It is essential that the project planning and design process include the identification of such designations, the reasons for the designation and an environmental impact assessment. This should identify any risks to the scheme. Without due consideration of this in the planning and design process there is potential for consequential delays, cost overruns and even scheme cancellation.

2.3.8. Health and safety

Ensuring health and safety is not only a statutory obligation, but is also seen as the citizen's duty under common law. Designs in the UK need to comply with regulations during all stages of design, construction, operation and decommissioning. The health and safety issues for each of these stages need to be assessed and either reduced or mitigated against before detailed design can start. A scheme should not be taken forward if the risks cannot be either removed, reduced or sufficient protection provided to the individuals interacting with the structure.

2.3.9. Political

Political risk reflects the influence of others to force changes in the way that a project is procured or the type of works proposed. This influence may come from a global, national, organisational or local perspective. The focus of political risk is largely influenced by the impact of past events and expectations for the future.

Some examples of political pressure are as follows.

- At an international level, there is political pressure to make improvements to our environment through treaties such as the Kyoto Agreement. With the withdrawal of the USA from the agreement, it is unclear whether or not the international political pressure to comply will be as strong.
- At a national level, the flooding in the UK during Easter 1998 prompted a review of flood defence and every flood event since, such as in autumn/winter 2000, has increased the pressure to make improvements in the management of floods.
- At a local level, there may be a strong opinion for a certain style of solution or the need to improve more than just primary objectives of the project. For example, there may be political pressure in a coastal town to recharge a beach to increase tourism even though a rock revetment along the seawall may provide the economic and engineering optimum solution.

All parties involved in design need to be quite clear about the political situation and how it may change in the future. For areas where change may occur, it is critical that sufficient contingency plans are developed to mitigate against political risks.

2.3.10. Societal

Societal risk is defined as the 'protection of human life'. The general public has an expectation that a structure will not endanger life. In discussions regarding acceptable number of fatalities, designers tend to not quantify but rather use broad statements such as 'protect life' or 'no risk of death'. However, failures of coastal and fluvial structures have led to fatalities and, therefore, designers should at least understand the magnitude of the risk being accepted and propose risk management plans to keep these risks as low as is practically possible.

The design can take two points of view when regarding societal risk:

- *individual risk* — a personal choice or situation, some of these risks are voluntary, such as smoking, while others are not, such as age or gender
- *total population risk* — the relation between frequency and the number of people suffering from a specific hazard.

In terms of probability of occurrence, the tolerated annual probability of being killed is regularly linked to accident statistics, which sets the probability to be approximately 10^{-4} (see Box 2.2). There is an argument, therefore, that the designer should at least ensure that his or her structure does not make the situation any more risky. The risk levels embedded in structural codes are certainly based on this argument.

For fluvial and coastal engineering projects in the UK, it is generally accepted that residual societal risk is controlled through warning, with either evacuation or limiting access to the structure at certain times. This process results in societal risk being considered more in qualitative terms and with the assumption that the structure does not contribute to the societal risk.

In comparison, the Netherlands has a more quantitative approach to societal risk, especially with regard to flood defence. This quantified approach is largely a function of the limited space in which to evacuate individuals from risk zones and the sheer scale of population that are at risk. The Dutch, therefore, have been more active in strengthening their structures to resist more extreme events than those used in the UK.

In summary, both the UK and the Netherlands have the primary objective to protect life. However, in the Netherlands the additional expense required to raise the standards of defence to protect individuals is justified by quantifying a fatality into a monetary term. The general consensus is that the value of life should be linked to the

---

Box 2.2. **Accepted annual probability of failure based on societal risk (CIRIA, 1976)**

During the development of the structural engineering codes, the accepted annual risk level in the UK for an individual was 10^{-4} (CIRIA, 1976) based on the probability of death by accident. The total risk to the population is a function of probability of death, population at risk and the nature of the activity associated with the structure.

The accepted annual probability of occurrence can be defined as:

$$P = \frac{10^{-4}}{n_r} \times K_s$$

where n_r is the number of people at risk

K_s is the social criterion factor (see table below).

Social criterion factor (CIRIA, 1976)

Nature of the structure	Social criterion factor, K_s
Place of public assembly, dams	0·005
Domestic, office or trade and industry	0·05
Bridges	0·5
Towers, masts, offshore structures	5·0

---

local economics for the particular area. Using this definition sets a value for life that is lower in a poorer country than a richer country; this may be difficult for an individual to justify morally, that in monetary terms they are worth more than another. However, linking to a national economic indicator ensures that all economic comparisons are within the same framework (i.e. capital costs in a poorer country are linked to the wealth that the country generates).

Some of the main techniques to quantify the value of life are based on:

- occupational-fatality risk and wages (Hammit *et al.*, 2000)
- present value of Net National Product per head (Vrijling and van der Gelder, 1999)

- income, risk, education and age (Bowland and Derghin, 1998)
- life quality index (Nathwani *et al.*, 1997).

Societal risk has been discussed at length within the industry. The key guidance to emerge from such discussions is that risk to life should be considered and documented. However, there is disagreement as to whether or not a qualitative review with management plans to remove risk are more appropriate and useful than a detailed quantification of the risk either in terms of an annual probability or in monetary terms. In the UK, quantification of the risk may be appropriate where reliance on the structure's performance to prevent fatalities is critical.

2.4.   STAKEHOLDERS AND RISK OWNERS

Stakeholder interests will most likely cover one of the generic risks described in the last section. Some of these stakeholders will be responsible for specific risks and hence, as risk owners, while others will have only an influence on risks.

The key stakeholders in a coastal and fluvial project are:

- funders
- designers
- insurers
- environmental, heritage and other pressure groups
- the general public.

The key risk owners will be the funder and the designer as these are the parties liable if risk occurs. The majority of the other stakeholders will not own the risk as such but will have influence on the risk owner; in some cases this influence will formalise itself as a statutory power (e.g. with English Nature and English Heritage).

2.5.   DEVELOPMENT OF DESIGN TEAMS

This chapter has progressed from discussing how a risk or a concern has certain levels of tolerability to the wider issues of multiple risks with individual owners, who may be influenced by a wider group of stakeholders. The next step is to discuss how these stakeholders operate as a design team.

First let us consider the mechanics of the design process and how design teams operate. Louis Bucciarelli, who studied design teams in the USA, concluded that *all* stakeholders are an integrated part of the design team:

> The realisation that design is a social process, that alternative designs are possible and that a design's quality is as much a question of culture and context as it is of a thing in itself or of the dictates of science and market forces.
>
> (Bucciarelli, 1994, p. 199)

This gives a picture of the design process as an interaction between various elements of society and the world around them. Each stakeholder decision is based on the specific needs of their own discipline-determined world such that they construct their own ideas of what is acceptable and unacceptable. In the end though, the structure designed and constructed is the result of influence and compromise rather than just pure technical analysis. With this social interaction, the final result may be less than perfect because:

> the object is not one thing to all participants. Each individual's perspective and interests are rooted in his or her special expertise and responsibilities. Designing is a process of bringing coherence to these perspectives and interests, fixing them in the artefact. Participants work to bring their efforts into harmony through negotiation.
>
> (Bucciarelli, 1994, p. 197)

A key component of effective design is communication and it is therefore important to ensure that all of the stakeholders are contacted as early as possible in the design process. A model of this process can be described as a communicative network (Figure 2.2) of all the individuals, groups or organisations involved in the design (primary network). Within their own groups and organisations, there will be subsidiary secondary and tertiary communicative networks informing the primary network of key issues that they need to consider. Without this communication structure, individuals will operate in isolation and as a result, the final solution may not be desirable to all parties. The cooperation of all stakeholders is important to the success of the project, since even if all aspects cannot be addressed then at least all parties have had an opportunity to influence the process.

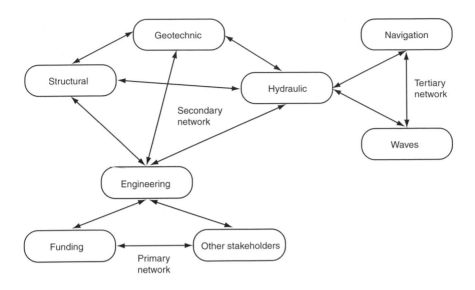

Figure 2.2. Example of communicative networks

2.6.   RISK ASSESSMENT AND MANAGEMENT

Each stakeholder will undertake risk assessments of the potential risks appropriate to themselves and start to define what is unacceptable and acceptable. Formal risk assessment and management approaches are now becoming more established within the engineering community. The key elements considered are:

- current risk level — probability and consequence
- risk management strategy
- residual or accepted risk level — probability and consequence.

There are several approaches available to manage risk. The general approach is to attempt to remove or isolate the source of the risk by breaking the pathway between the source and the result (i.e. showing the importance of sketching fault and event trees, as shown by examples given later in the book). The generic approaches to risk management are summarised below.

(*a*) *Removal of risk*. Removal of risk means avoiding the possibility of a risk arising by removing the hazard.

(*b*) *Reduction of risk*. Most options in coastal and fluvial engineering will involve risk reduction rather than complete removal. Typical approaches include:

(*i*)   reduce the likelihood that the risk is realised
(*ii*) reduce the consequences
(*iii*) reduce both the likelihood and the consequences.

Even if the risk is reduced so that it becomes insignificant, it may still occur and the designer needs to at least acknowledge that it remains, even if the risk is not quantified.

(*c*) *Contingency planning*. With risks reduced or even increased, the manager of the structure needs to be active in his or her risk management. Contingency plans are essential where there is a high level of uncertainty and should allow for different responses depending on events in the future. Allowing various options to be considered avoids being committed to a particular strategy, scheme or design too early, and may involve choosing versatile options that can be adapted to future conditions/events that are uncertain at present. Monitoring should be an integral part of these approaches and may be used to trigger certain actions or responses according to a pre-planned strategy.

(*d*) *Acceptance of risk*. A risk is accepted actively if there is a positive decision to allow the risk. Risks may also be accepted passively, either because they have not been identified or no action is taken by default. Passive risk should be minimised, a core aim of the risk identification, assessment and management process.

(*e*) *Transfer the risk*. It may be possible to transfer the risk to another party. An example of this is taking out an appropriate insurance policy or deciding to ring-fence funds to be made available if the risk actually occurs. Some elements of the design will not be insurable, so the designer must make a decision if the risk should be accepted or an option is sought to remove or reduce the risk.

The management of the risk assessment process is now more automated with the development of proprietary software. There are a number of 'off the shelf' software packages available, but the reader is directed to the latest software designed specifically for the construction industry through the Construction Industry Research and Information Association (CIRIA), which was developed by a consortium made up of HR Wallingford, the University of Bristol, Robert McAlpine and Currie & Brown (CIRIA, 2000). It provides a management tool that takes the user through the process risk assessment step by step. These tools aid the risk manager by collating data automatically, but the actual risk assessment process is not — and should never be — automated.

These systems deal directly with the definition of actual risks and setting out a management plan to remove or reduce them. However, they fail to deal formerly with the issue of tolerability or to visualise the social aspects of design. Both are inherent in the risk assessment process. Therefore, it is one of the primary objectives of this book to conceptualise how a designer can manage these multiple risks and concerns such that an optimum solution can be developed that satisfies all stakeholders that their position is at least tolerable even if not broadly acceptable.

2.7. THE CONCEPT OF THE ACCEPTABLE RISK BUBBLE TO MANAGE MULTIPLE RISKS

Identifying a generalised structure for managing multiple risks is difficult because each project is specific and rarely falls into the same categories. However, through the research undertaken for this book, the concept of an Acceptable Risk Bubble was developed to demonstrate and visualise both the concept of tolerability and the interaction of stakeholders during the design process.

The Acceptable Risk Bubble is a conceptual model of all the risks involved in the project, with the centre of the bubble being total acceptance and the outer surface being unacceptable (Figure 2.3). Each virtual axis represents a particular risk or concern and as one moves along that axis towards the outer surface, the tolerability of the risk decreases until it becomes unacceptable. As stated previously, by the HSE (1999), no risk or concern is totally acceptable and, therefore, the assumption is made that risks or concerns will fall into a broadly acceptable region, represented by the core of the bubble (lying within an 'inner skin').

The dimensions and units of each virtual axis are constrained by the risk owners. They can define their own levels of tolerability by

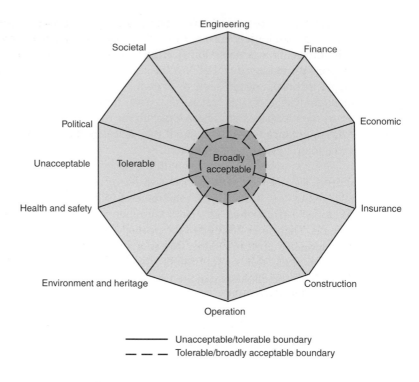

Figure 2.3. The Acceptable Risk Bubble

using either qualitative or quantitative terms. For example, a funder may say that they want to double the return on their investment but they would tolerate just breaking even while it would be unacceptable for there to be a negative return in their investment. This qualitative description could then be converted into an economic measure such as a rate of return or BCR.

The benefit of plotting all of the risks together is that each stakeholder can understand the tolerance levels of the other stakeholders and aid communication between stakeholders. This concept then allows the designer to look at the key issues that need to be considered and begin to get a feel for the bigger picture in terms of setting appropriate risk levels.

However, it is noteworthy that some criteria may not be independent and may even be sub-sets of each other and, as a consequence, they should not be necessarily given equal weighting. This issue can be incorporated into the bubble by scaling each axis individually. In practice, therefore, each axis will not have the same proportions for each criterion, causing the bubble to assume an irregular shape.

2.8. APPLICATION OF THE ACCEPTABLE RISK BUBBLE AS A MANAGEMENT TOOL

Conceptually, the Acceptable Risk Bubble is a powerful tool to help the designer manage multiple risks efficiently. In practice, this thought process might be used by a number of designers but there is very little formal information available on what is involved. The aim of this section is to highlight the key steps that the designer should consider when setting up or amending the Acceptable Risk Bubble for a specific project.

The main steps are:

- identify stakeholders and risk owners
- brainstorm risks/concerns and rank them
- select top level concerns
- define broadly acceptable, mid-tolerable and unacceptable levels
- compare risks using the Acceptable Risk Bubble.

Each step is described below, setting out the practicalities of the Acceptable Risk Bubble approach. An example has been provided in Box 2.3 to illustrate the key points described.

2.8.1. Identify stakeholders and risk owners

It is clearly important to identify all stakeholders and risk owners as early on in the process as possible, so that all risks and concerns are collated. This process can be achieved using a number of methods, but is generally done by the client and designer through identification of the key stakeholders. Public consultation exercises are used to identify other potential stakeholders who can then voice their own concerns and aspirations. The internet is helping to improve the communication process by providing access to a wider audience. See Section 2.5 for a list of the key stakeholders and risk owners that should be considered.

Box 2.3. Production of the Acceptable Risk Bubble for the Dawlish seawall

The Dawlish seawall (Case study 3) protects the main southwest railway line in the UK and is maintained by Railtrack (the company responsible for rail infrastructure in the UK). Two schemes were analysed using the 'bubble' concept: refurbishment of the seawall toe {Project (Seawall)} and nourishment of the beach {Project (Beach)}. The key stages of the process were as follows.

- *Identify stakeholders and brainstorm risks, rank them and select top level risks* — the key risks were identified as those presented in the diagram.
- *Define broadly acceptable, tolerable and unacceptable levels of risk* — a table was drawn up in which the risk threshold levels for each risk or concern were defined qualitatively; see example table below for one of the risks.

Extract from the tabulation of risks for the Acceptable Risk Bubble

Risk/concern	Broadly acceptable	Mid-tolerable	Unacceptable
Operation	No delays due to wave-induced overtopping or breaching of the seawall	Infrequent delays due to wave-induced overtopping	Large delays due to seawall breaching causing loss of track support material

- *Score projects against the Acceptable Risk Bubble and plot results* — the figure opposite shows that improving the beach {Project (Beach)} seems a stronger option than the refurbishment of the seawall toe {Project (Seawall)} when compared with the majority of risk criteria. However, on economic grounds, the BCR was found to be too low to make the scheme feasible. Therefore, the client accepted the refurbishment of the seawall toe as the best option.

This example demonstrates a classic dilemma for the designer in terms of the extent to which it is worth optimising a technically

Box 2.3. continued

attractive scheme if it is likely to be rejected on cost or environmental grounds. This prompts the question: is it worth taking the 'risk' of investing further effort on such a scheme in order to achieve a better balance of stakeholder aspirations?

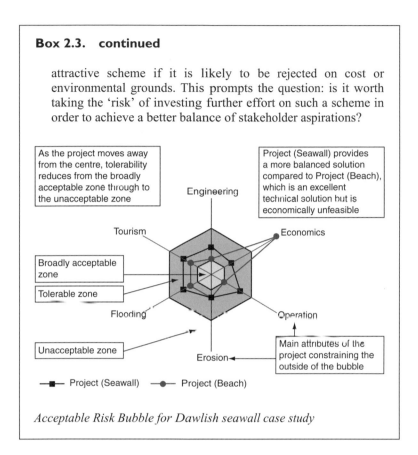

Acceptable Risk Bubble for Dawlish seawall case study

2.8.2. Brainstorm risks/concerns

Once all stakeholders have been identified, the next stage is to itemise all the potential risks and concerns that the designer will need to consider. It is becoming more common to bring key stakeholders together and run risk or value workshops to identify the key areas of concern (although in some projects the risks or concerns are generated generally by only the client and designer using data collected through the public consultation exercise). Whichever approach is adopted is not as important as ensuring that the list is communicated back to all stakeholders for further comment. This process is an iterative one of refining risks and concerns plus clarification that all stakeholders have been contacted. See Section 2.3 for a list of the generic risk areas that should be considered.

2.8.3. Select top level risks/concerns

The risk assessment process will often generate a long list of risks and concerns, some of which will be negligible in the context of the whole project. To make the exercise manageable, therefore, the designer needs to filter the list and rank risks in order of criticality. This ranking process will be based on discussions with the client and other stakeholders. As a rule of thumb, the top ten risks should be taken onto the next step but the full list of risks and concerns should be kept so that the ranking can be re-evaluated as the project progresses.

2.8.4. Define broadly acceptable, mid-tolerable and unacceptable levels

As discussed in Section 2.3, the designer needs to have a clear understanding of how the profile of the value curve changes as the value (i.e. risk or concern) increases. In practical terms, the designer will ask stakeholders to define what they perceive to be broadly acceptable through to that which is unacceptable. In practical terms, this process can be undertaken by amplifying a table with qualitative or quantitative information (see Box 2.3).

At the start of the project, stakeholders will, in general, only be able to categorise risks as being either broadly acceptable or unacceptable. As the project develops then some of the risks that were unacceptable will start to become tolerable as the stakeholder is provided with more information and also as the stakeholder starts to compromise with other stakeholders to obtain a solution. This process is discussed in more detail in Chapter 3.

2.8.5. Compare risks using the Acceptable Risk Bubble

Once a number of projects have been generated, then the designer can start to score the project against each axis and plot the result on the Acceptable Risk Bubble diagram (see Box 2.3). The diagram provides a visual tool to allow the designer to spot areas that may require further information to be collected or further clarification on that value curve.

The shape produced by plotting the results is very likely to change as the project develops and the definition of risk levels change, an issue that is discussed in more detail in Chapter 3.

2.9. SUMMARY: KEY ASPECTS OF SETTING UP A RISK FRAMEWORK

The key aspects of setting up a risk framework identified in this chapter are as follows.

- Each project will have its own set of risks and concerns. The needs of the project and the influence of all the stakeholders will largely define what risks are more important than others.
- Tolerability of a risk can be categorised into three levels — broadly acceptable, tolerable and unacceptable. In design, the tolerability of each risk or concern should be considered very carefully, together with its sensitivity to change.
- The designer should note that design is not merely an analytical process of technical analysis but is also a social process that involves both negotiation and compromise.
- The concept of the Acceptable Risk Bubble allows the designer to set out the design criteria and visualise the impact of key risks when considering a number of solutions. It also focuses the designer's mind of areas where further investigation is required. The process also promotes consensus building among stakeholders, which will be an important part of the project's success.
- The Acceptable Risk Bubble provides an opportunity to represent both qualitative and quantitative risks in one model.
- The Acceptable Risk Bubble complements and helps to present the results of the current risk assessment process.

Life-cycle management of physical schemes using the risk framework

3

3. Life-cycle management of physical schemes using the risk framework

This chapter is focused on the generalities of the system or structure life-cycle and how the concept of multi-attribute risk assessment discussed in the previous chapter is managed and evolves throughout this process.

3.1. OVERVIEW OF THE SCHEME LIFE-CYCLE

The way in which a scheme is planned depends on the final product. Some structures are familiar to a particular engineer and the process involves variations on familiar themes. An experienced design team will approach this task with confidence based on proven practice. On the other hand, the approach to an innovative structure may cause the designer to be far more cautious and to be aware that prior experience alone may not be sufficient.

Each type of engineering, from systems through to mechanical and civil engineering, have their own favourite process in developing schemes. However, a top-down approach is generally adopted where the constraints of the problem are defined at the start, followed by a generation of a number of solutions before the best solution is selected based on these constraints. This iterative process optimises the design to reap the maximum benefits.

For coastal and fluvial engineering, it is perhaps more appropriate to think in terms of continuous management of an infrastructure system or set of structures, as it is unusual for an individual coast or river structure to be removed once the project is complete. When the life of an asset does come to an end, there is generally a major intervention in the form of a new project and the asset management process starts again. Therefore, the principal stages of the scheme's life-cycle in coastal and fluvial engineering projects are (Figure 3.1):

- Stage 1: Need identification
- Stage 2: Functional analysis
- Stage 3: Generation of alternative solutions
- Stage 4: Comparison and selection
- Stage 5: Final design and detailing

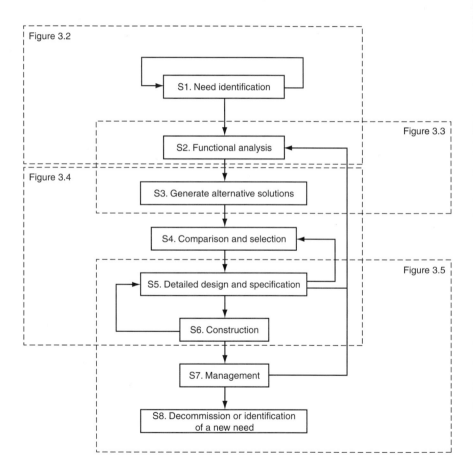

Figure 3.1. The generic scheme life-cycle used in coastal and fluvial engineering projects

- Stage 6: Construction
- Stage 7: Management
- Stage 8: Decommission or identification of a new need.

As the scheme passes through each stage (either conceptually or physically), new information will be obtained and processed that needs to be fed back into the Acceptable Risk Bubble (see Chapter 2) ensuring that the final solution is acceptable to the majority of stakeholders. Therefore, communication between the designer or asset manager and the stakeholders throughout this process is critical to the project success. This is an important point that some designers and managers may sometimes neglect due to commercial pressures to complete a design. In the end, however, there will always be a time or cost delay if a consensus has not been established correctly. The following sections take the reader through each step in the asset life-cycle and highlight the main issues that need to be considered when managing and integrating multi-attribute risks through the project process.

3.2. STAGE 1: IDENTIFICATION OF A NEED

An individual, normally the person or organisation that becomes the client/promoter, will perceive that a certain need exists (e.g. through public pressure on a local maritime authority to reduce flooding or coastal erosion, or preferably they will have planned in advance to address this need and have been monitoring it). The solution to this need may be the construction of a port or the upgrade of a flood defence, but at the initial stage this will still only be a perception. The process and issues of moving from identifying a need to functional analysis of the scheme are illustrated in Figure 3.2.

Once a need has been perceived, a designer (this may still be the client or external consultant) will be employed to investigate further and will be commissioned to:

- collect data, such as:
  - hydraulic conditions
  - planning policies
  - costs/budgets
  - identify stakeholders
  - site investigation

- analyse data, undertaking tasks such as:
  - historical analysis
  - desktop studies
  - numerical modelling
  - consultation with stakeholders
- identify design constraints in terms of:
  - physical boundaries (both on land and in the water)
  - funding
  - economic considerations
  - time restrictions

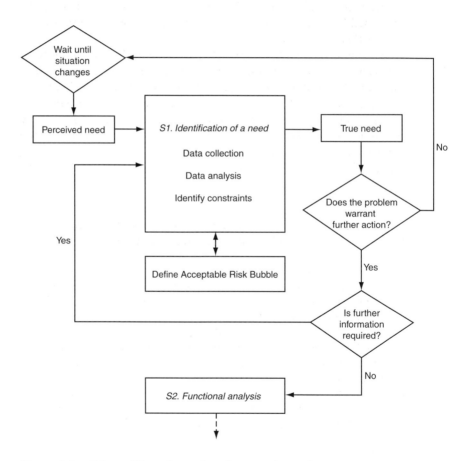

Figure 3.2.   Scheme life-cycle — identification of a need

- ○ environmental issues
- ○ political and social limitations.

The key aspects during this stage are to identify all stakeholders and ensure all relevant data are collected, analysed and then finally communicated to the stakeholders. Failure to undertake these steps may generate problems in the latter stages of the project, such as stakeholders objecting to certain designs or the introduction of new design data late on in the project that may have large impacts on cost and time. One such example is the Clacton case study (Case study 1) where late design data regarding the location of a natural heritage area changed the original rock groyne design to the construction of a detached breakwater. This change, fortunately, had little impact on the final cost of the design.

In the need identification of coastal and fluvial engineering projects, there seems to be a move towards a systems approach to defining the constraints of the problem. Consultation with clients and designers showed that the use of system diagrams was effective in setting boundary conditions/constraints. For example, it is very important in coastal and fluvial engineering projects to select the appropriate physical boundaries such that there is no double counting of benefits arising from the existence of adjacent assets (see Box 3.1).

Box 3.1. Example of a system diagram to define the physical boundaries of a problem

In coastal and flood defence, funders have to make informed decisions about where to do work and where to wait. In the UK, DEFRA use a priority score that sets out which schemes require funding first. This priority assessment is based primarily on economic assessments, but can also be influenced by environmental or political pressures. In other situations, the decision to fund works may be based on a wide range of reasons such as a private landowner wishing to make individual changes.

  Clacton (Case study 1) is an interesting case study because the natural flood zone is split into two parts, the study area and the adjacent flood area protected by the adjacent coastline west of the study area between Jaywick and Colne Point. A secondary flood defence comprising an earth embankment that extends perpendicular

---

Box 3.1. continued

to the coastline separates the study area from the 'other' flood area. If one of the two frontages fails then the secondary defence protects the area on the opposite side of the secondary defence. During the selection of which frontage to consider first, this secondary defence was assessed to ensure that if flooding did occur on one side then it would be contained within its own flood area. This system can be described as running in parallel because failure of one component does not impact on the other (see figure below).

However, the Clacton frontage is split into three main bays (i.e. the seawall between each fish tail groyne). If the seawall in one of the bays fails then the whole system will fail because they are connected to the same flood zone and therefore failure has the same consequence. This system can be described as running in series because it only takes one of the defences to fail to cause an impact on the whole flood zone (see figure below).

When designing a system that runs in series, there is little value in designing all of the defence lengths to different risk levels as it is the weakest link in the series that will cause failure. Funders, such as DEFRA, will wish to fund a scheme that encompasses protection of all of the elements that may lead to failure. Care needs to be taken by the designer to ensure that the system, its links and the consequences of a partial/total failure are properly understood. This will ensure that a design issue is not overlooked and any weak links in a design are identified.

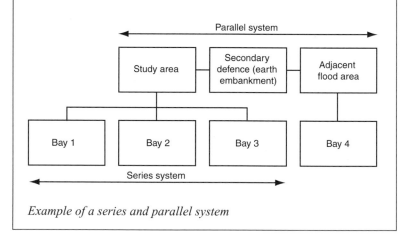

Example of a series and parallel system

If the designer systematically goes through the process of identifying stakeholders, data collection and identifying constraints then it should be relatively easy to develop a risk framework using the Acceptable Risk Bubble concept. As discussed in Chapter 2, when considering risks and their levels of tolerability early on in the project, the majority of risk owners will place risks in the unacceptable box and only put risks that they consider highly improbable or that they consider will have minimal consequence in the broadly acceptable box. It is perceived that the tolerable box would be left blank at this stage, so that each risk owner or stakeholder has a negotiating position later on in the project cycle. The bubble diagram at this stage will be only a first attempt in setting the parameters of the project ready for the designer to start considering the functional analysis of the project.

Developing an Acceptable Risk Bubble for a specific project will involve standard risk assessment methods, such as producing risk registers through individual consultation, surveys and risk workshops. This assessment can be considered as a two stage process (Box 3.2):

- *broad-brush* — define the risk and establish its order of magnitude
- *detailed* — further analysis to gain a better understanding and quantification of key risks.

A summary of the key risk assessment techniques is given in Appendix 1. There is a range of further literature available; for coastal and fluvial engineering, the DEFRA *Flood and coastal project appraisal guidance — approaches to risk* (MAFF, 2000) is a useful starting point.

The overall process of identifying a need is iterative with the designer, perhaps developing the bubble only after a number of passes, satisfies him or herself that the need requires further action and that no further information is required to move on. If the need is not sufficient then the designer should advise the client to wait until a time arises when there is a requirement to address the need. However, setting up the Acceptable Risk Bubble allows the change in the need to be monitored more efficiently by contacting stakeholders periodically and updating information. This process has been observed in many coastal communities in the UK through the establishment of coastal forums and flood defence committees.

Box 3.2. Example of broad brush and detailed risk assessment (Dawlish seawall)

In the analysis of the Dawlish case study (see Case study 3), a broad brush analysis was undertaken to determine the likely path between train delay and cause. A more detailed risk assessment of the causes was then undertaken to prioritise those that needed the most attention in designing a solution (see the table below). In this case, breaching was a major risk compared to the others, so the designer decided to focus on this aspect rather than trying to reduce localised overtopping.

Back analysis of past delays on railway line

Event	No. of events	Approx. duration of event: min	Approx. delay to trains: min
Catastrophic failure of the seawall	2	63 000	10 500 000
Loss of track support material	1	42 720	72 091
Wave-induced overtopping	15	8000	3000
Landslide material landing on track	1	150	270
Error in safety checks	1	90	100

Box 3.2. continued

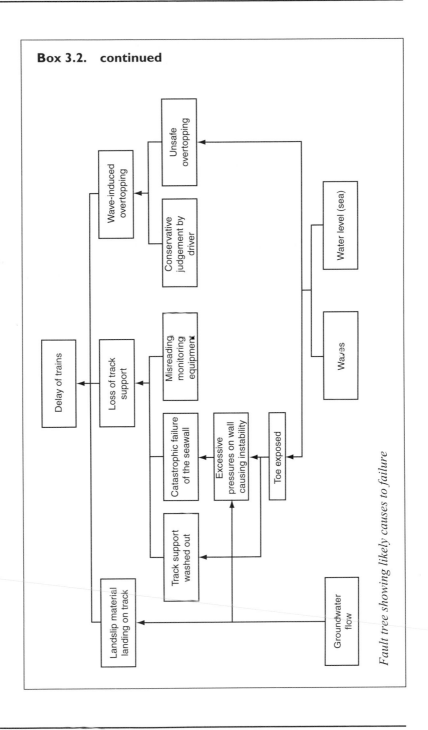

Fault tree showing likely causes to failure

3.3   STAGE 2: FUNCTIONAL ANALYSIS

This is the formal process of setting the Acceptable Risk Bubble into a technical specification, which will be used to generate solutions to the need and allow comparison. The process and issues of moving from functional analysis of the scheme to generating alternative solutions is described below and also presented as a flow diagram (see Figure 3.3).

The technical specification developed within the functional analysis stage must, as a minimum, match all the risks below an unacceptable level as defined using the Acceptable Risk Bubble. If part of the functional analysis addresses an area that is not covered by the Acceptable Risk Bubble then the designer must re-evaluate the particular specification to check that it is actually required and to identify which stakeholder is responsible or identify a new stakeholder to set constraints. This whole process continues to focus

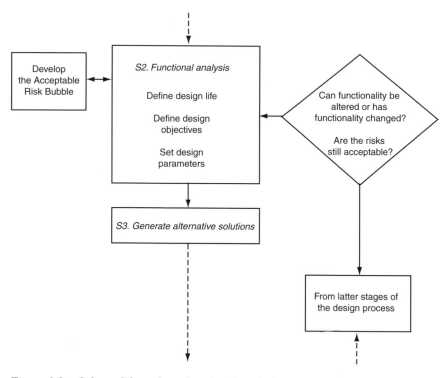

Figure 3.3.   Scheme life-cycle — functional analysis

on the consensus building achieved during the need identification stage.

As it is a focal point of the project, the designer may return to the functional analysis if the technical specification cannot be satisfied. The solution is to return to the 'bubble' and start negotiation with all the stakeholders, who may then start to allow some risks or concerns to move into the tolerable zone, as their point of view has to be compromised to reach a collective solution. Note that a change in functional analysis may also come from a change in the stakeholders' position rather than from a design not meeting the current specification (e.g. a change in Government policy).

If the overall process of defining the functionality of the scheme breaks down, the project should be stalled until a collective agreement can be found. However, in some cases where a stakeholder has no real political influence or their arguments are not relevant to the collective, then their stance may be considered and noted but not taken forward so that the project can continue. However, this will largely depend on the particular situation, and the designer should try to ensure that the views of all stakeholders are considered. With careful consultation and discussion, such a situation should rarely arise.

The key elements that the designer needs to address generically at the functional analysis stage are as follows.

- *Defining the design life* — this issue is discussed in more detail in Chapter 4.
- *Setting acceptable levels of failure* — this issue is discussed in more detail in Chapter 4.
- *Defining appropriate design events* — this issue is discussed in more detail in Chapter 4.
- *Defining potential changes to the design functionality* — this issue covers such points as climate change and other potential changes that may impact on the project. How a designer identifies and manages these potential changes are discussed in more detail in Chapter 4.
- *Financing* — how a project is going to be financed has a large influence on the project; funders need to be convinced that their investment will be used efficiently.
- *Economic return* — the normal situation is that the client wishes the benefit of the solution to outweigh the whole-life cost (capital and operational expenditure) in order to produce the solution. The

key economic indicators that can be used to define a target limit are:

○ BCR
○ net present value (NPV)
○ internal rate of return (IRR).

The most common indicator used in the UK for government-funded schemes is BCR, which is expected to be greater than unity over a period of 50 years (although for funding to be awarded, a BCR higher than this is generally required). Government also dictates that any economic assessment should use a discount factor of 6% and inflation should not be considered in the calculation. With other private developers, the NPV or the IRR of their investment may be considered. Generally, a private developer will look for shorter investment periods in the range of 15 to 20 years. Economic appraisal lifetimes are discussed further in Chapter 4.

- *Environmental impact* — in the UK, the broadly acceptable risk level for a scheme's environmental impact is that no damage occurs. In some cases, damage may be accepted when an environmental benefit is created somewhere else or the risk is reduced to as low as practically possible, but this will largely depend on the nature and importance of the environmental asset. Therefore, risks are generally defined qualitatively rather than quantitatively.

- *Specifying finishes and aesthetics* — the aesthetics of a scheme may be important to fit in with its local environment or heritage and there may be a requirement to use specific local materials. Care needs to be taken not to reduce the creative process at this stage so the designer should be encouraged to keep this as flexible as possible as these issues can be dealt with more effectively during the detailed design. There will be times when certain items will need to be specified earlier on because of issues such as material availability (e.g. if a certain size of rock is all that is available then there is little merit in designing a structure that requires larger rock).

- *Other miscellaneous items* — there will be a number of other items that will be part of the project development, however, as discussed above, it is important to let the Acceptable Risk Bubble aid the design process instead of constraining it by being too specific at an early stage. It is more about understanding the concerns of stakeholders in a broader sense so that the designer still has some freedom to be creative.

3.4. STAGE 3: GENERATION OF ALTERNATIVE SOLUTIONS

The designer has defined the true need, collated the views of all of the stakeholders into different risk levels and then set out a list of criteria on which to measure the performance of the design. For a given need and functionality, there will always be a number of different solutions, which will then be compared to find the optimum solution that will be designed in detail (Figure 3.4). The solutions will be specific to the need, but all of the solutions will fall into one of the following categories:

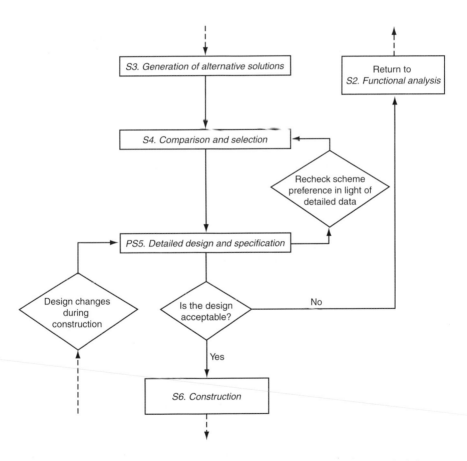

Figure 3.4. Scheme life-cycle — generation of solutions through to detailed design

- do nothing
- do minimum
- minor investment
- major investment.

The 'do nothing' assessment is an important solution because it is the benchmark solution against which the rest of the solutions will be compared and the best selected. The process of generating solutions will make use of previous designs, personal preference, brainstorming and the use of generic lists. The process should not constrain creativity and there should be little concern over whether options are fully acceptable or not. The most important point is to produce a comprehensive list of options that covers the full range of categories described above. Again, it is useful for the designer to undertake this process with the key stakeholders so that their ideas are integrated into the process.

3.5.   STAGE 4: COMPARISON AND SELECTION

Once the designer has collated the range of alternative solutions, the next task is to compare the options against the 'do nothing' assessment and select the best option that also satisfies the functional analysis criteria. If a number of alternatives have been generated, then it is useful to initially undertake a broad-brush review of all the options and discard those that are not sustainable. This broad-brush approach is generally undertaken by considering broad issues, for example, if the option is impractical to construct or will result in substantial damage to the environment. The designer should also start to think about how the risks involved may change through the lifetime of the project, from construction to decommissioning.

Once the options have been reduced to a more manageable number then a more rigorous qualitative review can look at the pros and cons of each option, broken down into a number of categories and considering the risks associated with them. The categories should be defined by the functional analysis undertaken previously. Some examples of different approaches to comparing and selecting options are provided in Case studies 1, 2 and 3.

By the end of this process, the designer should have a limited number of different options that satisfy all of the functional analysis. Selection of the design that will be carried forward to detailed design should be the design that is the most acceptable as defined by the Acceptable Risk Bubble. A number of other selection techniques exist that can complement the bubble approach, such as a weighted multi-criteria analysis where the performance of specific attributes are weighted more favourable than others (see Appendix 1). As always, there is also an opportunity to negotiate once again with stakeholders regarding certain issues to improve the acceptability of the selected design, remembering that any changes must pass back through the functional analysis.

3.6.   STAGE 5: DETAILED DESIGN AND SPECIFICATION

The fundamentals of detailed coastal and fluvial design are discussed in Chapter 4, but the broader issues that need to be considered are discussed in this section.

At the detailed design stage, the designer has selected the best solution that can satisfy all of the functional analysis that has been processed through the acceptable risk bubble. Each project will be different in terms of the detailed design approach, but, as a designer, the main issues that need to be considered in detail are:

- detailed data collection, such as specific site investigations
- detailed modelling and analysis of all the design risks
- material specification
- type of contract to be used and method of procurement to be used
- planning requirements
- construction phasing
- monitoring and maintenance plans
- disposal options.

The designer will have a suite of design tools available and will select the best tools for that particular project; so in some projects it may be economically viable to construct a physical model while use of an empirical formula may be sufficient for a smaller scheme. It is essentially up to the designer to make a judgement on the level of detail required. This highlights the importance of employing well-trained engineers and design teams who have a wide experience in the work that is being asked of them. They also must have a good understanding of the different approaches to design from the three main disciplines of structures, geotechnics and hydraulics. In particular, there is a need to understand properly the differences between disciplines in setting risk levels.

During this detailed design and specification process, it may become apparent that there is some flexibility for optimisation of the design to save cost or time of completing the project, or in some cases the required standard may not be able to be satisfied. Once again the opportunity arises for the designer to negotiate with the required stakeholders and modify the Acceptable Risk Bubble so that the functionality can be changed accordingly. The establishment of the risk framework and opening up the communication lines between stakeholders described in Chapter 2 is clearly important to make sure changes or problems in the design are managed and accepted efficiently. However, the designer must always be conscious of not compromising safety to satisfy a particular need of a stakeholder (e.g. reduce cost). There is a professional responsibility to explain clearly if a particular stakeholder need is not achievable, especially on safety grounds, so that a safe solution is designed. This aspect of setting risk levels is discussed in more detail in Chapter 4.

3.7. STAGES 6 AND 7: CONSTRUCTION AND MANAGEMENT

Once the project enters the construction phase, the Acceptable Risk Bubble lies dormant (other than on the economic/financial axis) unless the contractor proposes an alternative solution or additional information is identified that leads the designer to return to the detailed design stage (Figure 3.5). However, once the scheme is constructed, the project will require some level of monitoring and maintenance to be undertaken at periodic intervals. Whenever monitoring or maintenance is undertaken, the operator of the structure should check that the functional analysis, and hence the form of the Acceptable Risk Bubble of the scheme, is still applicable; from this re-evaluation may emerge better ways to manage the asset and the underlying issues for which the scheme was developed than those proposed when the scheme was first envisaged.

The key difficulties that may arise during construction and maintenance are highlighted in the construction risk manuals on both coastal and fluvial engineering (Simm and Cruickshank, 1998; Morris and Simm, 2000) for a number of different structures, but the main issues to be considered are:

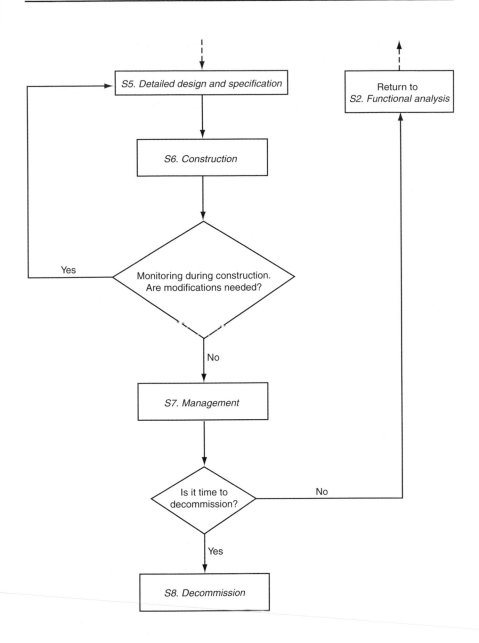

Figure 3.5. Scheme life-cycle — construction and management

- weather
- ground conditions
- site access
- buildability
- unrealistic specifications
- poor quality of materials and workmanship
- health and safety
- equipment malfunctions
- funding difficulties.

With all these issues, it is important to have considered them before the construction stage and to put in place a design that addresses them adequately. It is also important to have input from experienced contractors from as early a stage as possible, preferable during the functional analysis. If the procurement route involves a contractor being selected late in the day, then it may be necessary to involve a contractor or range of contractors separately to provide the necessary advice at the earlier stage. The time-scale for contact and appointment of contractors in the project process will largely depend on the confidence of the client and the designer.

Management of the structure is important, as positive lessons can be learnt form collecting data on its performance for refining the maintenance strategy or designing new structures. By returning to review the functional analysis regularly, the designer or the manager should be kept in touch with the acceptable risk matrix to make sure that both are balanced and there has been no change in concerns over time.

3.8. STAGE 8: DECOMMISSION

The final stage is the point where the decision is made to decommission the asset and either start afresh or stop all involvement, depending on how the nature of the project has changed. This process may involve abandoning, managed retreat or construction of a new scheme. Therefore, it is not unusual for the decommissioning stage to be associated with the start of a new project with a newly perceived need.

3.9. SUMMARY: KEY ASPECTS OF INTEGRATING AND MANAGING MULTI-ATTRIBUTE RISKS WITHIN THE SCHEME LIFE-CYCLE

In terms of integrating the Acceptable Risk Bubble concept into the scheme cycle, the 'bubble' will continually refine itself as the project develops through negotiation and compromise. The key aspects that need to be considered at each stage are as follows.

- *Need identification* — the designer needs to clearly identify all risks or concerns and stakeholders in order to establish the Acceptable Risk Bubble framework.
- *Functional analysis* — the designer converts the Acceptable Risk Bubble into a technical performance specification in order to generate and compare solutions. This stage involves consideration of functionality and indicative standards for different design limits.
- *Comparison and selection* — schemes will be prioritised against the technical performance specification and in some cases this specification will need to be changed through consultation with the stakeholders.
- *Detailed design and specification* — the designer seeks to optimise risk levels and check the robustness of the risk level accepted.
- *Management* — the designer can use the Acceptable Risk Bubble framework to monitor realisation of risks and to make decisions regarding future management or whether to consider decommissioning.

Application of the risk framework on a specific engineering solution

4

4. Application of the risk framework on a specific engineering solution

The focus of this chapter is on the application of the risk framework to a specific coastal and fluvial engineering solution. The process described should not be regarded as a checklist but rather as the building blocks of a philosophy which can be extended by designers on a project by-project basis. Throughout the design process, the designer must always draw on personal experience as well as the experiences of other designers through the use of appropriate codes of practice, guidance manuals and personal contact.

This chapter starts by defining the specific uncertainties that the designer needs to consider closely during the design of coastal and fluvial structures. The chapter then goes on to discuss the complexities of how the three main design disciplines (hydraulics, geotechnics and structures) interact and where risk levels can be influenced. The final part of the chapter focuses on a generic process that designers can follow to aid in setting the appropriate design risk levels by considering current practice and highlighting where improvements could be made.

4.1. THE SPECIFIC UNCERTAINTIES INVOLVED WITH COASTAL AND FLUVIAL DESIGN

With all civil engineering projects, there is a degree of uncertainty attached to design loading conditions and their responses. This uncertainty is an integral part of using models of observed processes to make robust decisions about an unpredictable world.

The specific uncertainties in coastal and fluvial engineering can be generally categorised as the following:

- variation in the design tools used
- variation in hydraulic conditions
- variation in ground conditions
- reliability of materials' strength and durability.

4.1.1. Variation in the design tools used

The selection of design tools and methods is a project-specific decision and is based largely on the economic benefit of undertaking the analysis. This book does not intend to set out explicitly which design tools should be used and it is therefore important that the designer is suitably experienced to ensure the most appropriate technique is used. It is also important that all designers continue to keep up-to-date with design methods and learn from other engineers' successes and failures.

The main issue is that a wide range of design tools are available (Table 4.1) and the designer needs to understand clearly the uncertainty surrounding each design tool. This understanding needs to be communicated formally in reporting, so that the tolerances in the analysis are understood and appreciated by all parties. Recent examples of poor communication have been observed in the UK, where design water levels predicted by river numerical models were taken at 'face value' and no tolerance for model error was built into the design; the undesirable consequence was of economic assets being flooded during an event that was predicted by the model to be safe (Bye, 1998).

Designers can be placed in such a situation where the level of investment for design is limited and a design tool that provides less confidence may be selected. In these situations, the designer needs to negotiate for the most appropriate tool to be used or use engineering judgement to build into the design sufficient strength to mitigate against the uncertainty in the results produced by the design tool. This will change from project to project.

4.1.2. Variation in hydraulic conditions

Hydraulic conditions are, by nature, variable and the designer must deal with the complicated issue of designing for both normal operation conditions and extreme conditions. Therefore, by definition, the majority of hydraulic descriptors used currently are probabilistic in order to deal with the randomness of these conditions

Table 4.1. The range of generic design tools available to the coastal and fluvial designer

Generic design tool	Description
Engineering judgement	A number of design decisions are based purely on engineering experience where there is no formal calculation approach established. However, simple crosschecks are required to ensure that this judgement is well formed
Historical analysis	Analysis of previous performance in some cases provides sufficient information for a design decision if the confidence level is high regarding future temporal changes in the risks
Fundamental equations	Generally used to describe the physics of the problem, can be either used independently or built into numerical models
Empirical equations	Based on specific observations, care needs to be taken when applying these techniques outside the boundaries of the original analysis
Numerical modelling	Combination of fundamental and empirical equations; again care needs to be taken to understand the limitations of the modelling. With some numerical models, a longer timeframe can be analysed. There are different levels of sophistication available and the designer should take advice from experts as to the best level to adopt
Physical modelling	Two-dimensional and three-dimensional models are used to analyse specific locations and are generally used as optimisation tools or to design complicated details that numerical models cannot analyse with any certainty. The scaling of a physical model will be accurate for a number of key parameters but at a consequence of others, so care is needed that the important parameters are scaled correctly

over a defined time period (e.g. significant wave height is the average third of the highest wave heights). The designer needs to make appropriate assessments of hydraulic conditions under the appropriate range of conditions to be considered in the design of coastal and fluvial structures. The key issues to consider are as follows.

- *Uncertainty in measurements.* Assessments of hydraulic conditions are based on statistical interpolation of observed or model data to predict the occurrence of frequent to extreme events. In some cases, expert judgement will be used to produce hydraulic conditions for a specific site based on observations from neighbouring areas. Therefore, to control quality, the designer may employ a specialist to make such measurements. However, it is still important that the designer understands the uncertainty in these data (measured or synthesised) and mitigates against this uncertainty by making appropriate sensitivity tests of the design to check for robustness.
- *Changes in the physical geography of the site.* The physical geography of the site is variable in the sense that beaches are mobile, river systems evolve, land use practices change. The designer needs to be aware of what changes may occur on a daily, seasonal, annual and long-term basis in order to make rational decisions about appropriate design conditions. Note also that predictions made by models of physical changes will have their own level of uncertainty.
- *Future changes in climate.* The designer tends to design using events that have already happened and then makes suitable sensitivity checks to ensure that the design is sufficiently robust to withstand any reasonable expectations arising from climate change (e.g. sea level rise, increases in precipitation, and changes in storm patterns). As numerical models improve, we may see climate change models starting to integrate seamlessly into the design process. However, it is clearly important that the designer takes into consideration the potential impact of climate change when designing coastal and fluvial structures.

There are a number of sources of information on climate change that the designer can access. The most common source used in England and Wales for sea-level rise is prescribed by DEFRA (2000) in their project appraisal guidelines for flood and coastal defence works (Table 4.2). Subsequent global estimates have predicted lower rates of sea-level rise when more complex feedback effects have been taken into account in the climate modelling. Periodic reports are produced under the UK Climates Impacts Programme that provide up-to-date predictions.

Table 4.2. *Regional rates of relative sea-level rise prescribed by DEFRA (MAFF, 2000)*

Environment Agency region	Allowance
Anglian, Thames, Southern and North East (south of Flamborough Head)	6 mm per year
North West and North East (north of Flamborough Head)	4 mm per year
South West and Wales	5 mm per year

However, with the degree of uncertainty involved, DEFRA see no justification for the general adoption of either reduced or increased allowances for sea-level rise. The designer should not be complacent about climate change, however, and should most certainly keep up-to-date with latest discussions on the subject, so that an appropriate level of uncertainty can be incorporated into the design of each individual project.

4.1.3. Variation in ground conditions

Ground conditions are highly variable and the resultant uncertainty, along with associated hazards, can be considered within three main aspects of geotechnics.

- *Ground boundaries.* Correct identification of the stratigraphy of the ground is vital to avoid risk — for example, consider the presence of thin strata of weak material not properly identified through a site investigation. The choice of suitable ground investigation techniques is covered in a range of literature (e.g. BS 5930 (BSI, 1999)). The designer should assess suitable methods for investigating the ground based on the anticipated ground conditions and the nature of the structure and its interaction with the ground.
- *Ground properties.* The properties of ground material vary with a range of factors, so that classification of materials in the ground may not identify uniquely the properties appropriate for use in design. Measurement of properties can be direct (through specific in-situ or laboratory tests) or indirect (through empirical

relationships to in-situ or index laboratory tests). However, the applicability of these tests in determining ground properties can be affected by a number of factors and the designer should carefully assess the reliability of the results obtained before determining a design parameter.

- *Groundwater*. Groundwater levels can be highly variable through seasonal and tidal cycles and can be highly influenced by local precipitation and infiltration. Assessment of the characteristic groundwater profile for a structure needs to consider both the predicted worst likely groundwater condition as well as an assessment of the worst credible groundwater condition (for example, as a result of a burst water main). Design of appropriate drainage measures should consider all possible sources of water and their likely variation to ensure that an accurate assessment of the groundwater is made.

4.1.4. Reliability of material strength and durability

Coastal and fluvial structures generically operate in very hostile physical environments compared to other generic civil engineering structures. The success of the design performance relies on the reliability of a material selected in terms of both its strength and durability.

In terms of strength, recommended design (ultimate) stresses are well documented for a wide range of conventional materials. Strength data are generally based on testing samples and care needs to be exercised to ensure that the strength characteristics assumed are properly representative of the materials and their application. Typical strengths are quoted to a reliability level of 5% of samples failing to meet the prescribed strength.

A definition of durability is the resistance of the material which will maintain acceptable strength within the structure. The types of issues include erosion, abrasion, chemical corrosion and biodegradation. The designer needs to consider the durability of the material carefully in order to ensure that the material either remains in place with adequate strength, or a maintenance strategy is put in place to maintain the required strength over the structure's working life (Box 4.1). The risk of durability is a temporal risk (which is discussed later in this chapter).

Box 4.1. Durability of steel sheet piles at different exposure levels

For example, the exposure of a steel sheet pile cut off toe on a seawall will depend largely on the beach levels at the time, wave conditions and the tidal fluctuations. Durability of the steel sheet pile will depend on the level of exposure. In this example, the level of exposure can be split into four zones (see table below).

Durability of steel sheet piles — levels of exposure (from Thomas and Hall (1992))

Zone	Description	Deterioration of steel
Zone 1	Salt spray and intermittent wetting by rainwater	0·1 mm/year
Zone 2	Intermittent wetting and drying by seawater and occasional rain	0·3 mm/year
Zone 3	Cyclic wetting by seawater with tidal fluctuation	1·0–0·2 mm/year
Zone 4	Permanent submersion	0·03 mm/year

The durability of conventional materials are documented in a number of manuals or manufacturers may provide data on their own materials. The durability risk is difficult to predict in coastal and fluvial design, however, and it is important, therefore, that the designer builds up a personal database of durability rates from both technical data and experience.

The level of uncertainty in a material's properties will be based largely on its past performance in similar environments. Where a new material is being used there will be a higher level of uncertainty about the performance. Methods available to reduce this uncertainty generally involve undertaking a phase of testing either on-site or within the laboratory.

4.2. CURRENT DIFFERENCES IN THE MANAGEMENT OF UNCERTAINTY BETWEEN THE MAIN DESIGN DISCIPLINES

Uncertainty in coastal and fluvial engineering is discussed above, however, the designer also needs to be aware of how each of the main design disciplines deal with these uncertainties. The key focus areas for each design disciplines are:

- the hydraulic discipline focuses mainly on loading — waves, water levels, currents, flows and sediment movement
- the geotechnics discipline focuses mainly on strength and stiffness of ground materials, groundwater and the resulting stability of structures, slopes, etc.
- the structural discipline focuses on strength and durability of materials, load paths and stress distribution.

In setting risk levels, it is clearly important to understand the approach within each of the main design disciplines with reference to the management of uncertainty. Uncertainties and their management are discussed regularly within each individual discipline of the engineering community through a wide range of literature. This book does not presume to enter those detailed discussions, but rather seeks to generalise on the main points that relate specifically to coastal and fluvial engineering. The approaches to manage uncertainty within each discipline are discussed below and some general comments made on the overall process of communicating uncertainty to the layperson.

4.2.1. Hydraulics

The hydraulic designer will use a design approach which involves either deterministic design with sensitivity testing or probabilistic design. The latter is often used where there is greater uncertainty (or variability) in the input and output conditions of the design. Codes of practice do exist, but the discipline relies on the use of best guidance with no developed safety factors. Loading conditions and structural responses are taken more or less at 'face value', meaning that risk levels are more clearly communicated as the designer is working with first principles rather than using prescribed safety factors.

4.2.2. Geotechnics

The geotechnical designer generally uses a deterministic approach by selecting design parameters and processing them through well-developed codes of practice to obtain a final lump sum factor of safety. Guidance is provided on selecting factors of safety, but there has been little standardisation and the final choice should be based on engineering judgement. Where there is significant uncertainty in ground conditions, sensitivity testing is often applied to ensure the integrity of the design. The risk level is more transparent but described as a factor of safety based on the risk level of a prescribed event rather than a probability of failure.

The geotechnics discipline is now starting to embrace the use of partial safety factors and probabilistic design methods (Eurocode 7 (BSI, 1995); see Box 4.2). However, the use of probabilistic methods in determining ground properties, for example, requires 'comparable experience with ground properties to be taken into account for example by means of Bayesian Statistical methods' (Eurocode 7, section 2.4.3(6)). This level of experience is available from very few designers. The use of 'lump sum' factors of safety is seen to be more practical and allows the impact or sensitivity of certain parameters to be assessed more effectively by the designer.

4.2.3. Structures

The structural designer generally uses a deterministic approach by selecting design parameters factored by a partial safety factor and processing them through a well-developed code of practice. Probabilistic methods were, however, used to define the partial safety factors (Box 4.2). The focus is on the application of safety factors to the loading conditions rather than the structural response because, in general, the designer has more confidence in the response to a particular load rather than a particular load actually occurring. Some sensitivity testing may be done as a final check in the design process. The risk levels are built into the codes of practice and are not transparent to the designer, i.e. the designer accepts the in-built risk levels which permits the design process to be semi-automated.

4.2.4. Communicating uncertainty

Under all disciplines, to mitigate the risk of designing a weak structure, designers will increase the strength or resistance of the structural elements to provide an appropriate margin of safety in their design. This margin is generally a balance between economics and the uncertainty in the design's performance (i.e. to minimise cost but not at the cost of unacceptable safety). The terminology used traditionally by the designer to describe this process is to 'over design' or be 'conservative'. However, to the layperson, these traditional terms are often understood quite differently and can be misleading.

Box 4.2. Changes in the use of partial safety factors on loading conditions (continued opposite)

The common approach to dealing with uncertainty of input conditions (i.e. loading conditions) is to apply partial safety factors. In hydraulic engineering, it is unusual to apply a factor of safety although in some instances a margin of safety is applied at the end of the analysis, such as a 0·5 m freeboard on a fluvial embankment. The use of partial safety factors was developed initially through structural engineering of buildings based on existing test results of a number of different structures. The results of this work were reported in *Rationalisation of safety and serviceability factors in structural codes* (CIRIA, 1977), which focused on a probabilistic assessment of loading and structural response. The typical partial safety factors for load that came out of that work and that are recommended in current British codes of practice for structural design are shown in the first table opposite (from BS 8110 (BSI, 1997)).

Note that, in addition, partial safety factors for strength of materials are applied to the characteristic strength of structural materials.

The development of Eurocodes has rationalised combinations of loads to permanent (self weight, ground and groundwater) and variable actions on structures. These are then applied to three design cases which should be verified separately for each structure. Design Case A covers loss of static equilibrium (where the strength of structural material or ground is insufficient), Case B covers failure of the structure or structural elements and Case C covers failure in the ground. The second table opposite is taken from Eurocode 7 (geotechnical design) and includes the partial material factors applied to the characteristic values of ground material properties to determine the design values for each of the cases.

Typical partial safety factors recommended in BS 8110

Load combination

	Load type					
	Dead		Imposed		Earth and water pressure	Wind
	Adverse	Beneficial	Adverse	Beneficial		
1. Dead & imposed (& earth & water pressure)	1·4	1·0	1·6	0	1·4	—
2. Dead & wind (& earth & water pressure)	1·4	1·0	—	—	1·4	1·4
3. Dead & wind & imposed (& earth & water pressure)	1·2	1·2	1·2	1·2	1·2	1·2

Partial material factors — Eurocode 7

Case	Actions			Ground properties			
	Permanent		Variable	$\tan\phi$	c'	c_u	q_u*
	Unfavourable	Favourable	Unfavourable				
A	1·00	0·95	1·50	1·1	1·3	1·2	1·2
B	1·35	1·00	1·50	1·0	1·0	1·0	1·0
C	1·00	1·00	1·30	1·25	1·6	1·4	1·4

* Compressive strength of soil

A 'conservative' or 'over designed' solution may be thought by the lay person to mean that a design has more than adequate margins for uncertainty. Because the margins are more than adequate, this may appear to mean excess provision and unnecessary expense. But what the designer is seeking to achieve are adequate margins for uncertainty, which may be provided by factors of safety, statistical factors, selection of calculation method or approach to the interpretation of field data. The margins will be larger for projects where uncertainty is greater, and smaller where uncertainty is less.

A design where it is appropriate to apply larger margins for uncertainty should not be described as being 'over designed' or 'more conservative'. Instead the designer should be clear and say that the structure is properly designed with appropriate margins for uncertainty incorporated into the design.

4.2.5. Key approaches to managing uncertainty in the design of coastal and fluvial structures

Uncertainty in the design of coastal and fluvial structures can be managed generally by using four approaches:

- sensitivity testing of results
- probabilistic modelling of loading and response
- use a lump sum factor of safety
- apply partial factors of safety to key parameters.

These approaches can be applied to both the uncertainty surrounding the loading conditions themselves or by analysing the structural response to a set of loading conditions. Current design practice is to select a loading condition to design against and make the assumption that the response is equivalent in terms of its probability of occurrence as the loading event. However, in some cases this assumption is wrongly made especially where bivariant conditions (e.g. waves and water levels) exist. Box 4.3 sets out an example that shows the potential difference in undertaking the design analysis based on loading conditions or structural responses. It is clearly important that the designer understands the sensitivity of the response to the loading event, so that uncertainty can be managed appropriately.

Within the main design disciplines, the approach to managing uncertainty is most flexible with hydraulics where the designer is working with first principles rather than structured codes of practice as with geotechnics and structures (Table 4.3). A more consistent approach across each of the disciplines is perhaps required so that at least uncertainty is managed in a similar framework within the design

---

Box 4.3.   Calculating design conditions

Data from the West Bay case study (Case study 6) can be used to illustrate two different methods of considering the design criteria — i.e. in terms of the:

- loading
- response.

Loading
In this approach, the 'design event' is considered in terms of the return period of the loading conditions. The main assumption is that the return period of the loading conditions equals the return period of the response function (in this case overtopping).

At West Bay the loading conditions consist of wave heights and water levels, it was therefore necessary to consider the joint probability of the loading conditions. Traditionally, a simplified assumption of independence or complete dependence between the loading conditions would be made. However, this assumption can lead to the under or over design of structures, as often there is partial dependence between these two variables. More recently, joint probability methods have been developed that consider the extent of any dependence between loading conditions. The most simple of these methods (CIRIA, 1996) provides combinations of marginal return periods (wave heights and water levels) that each have the same joint probability of occurrence (i.e. a joint probability contour) for a specified level of dependence. The overtopping rate is evaluated for each of the combinations on the joint probability contour, and the worst case (highest overtopping rate) is obtained.

Response
In this approach the response function is calculated directly. This can be carried out using a univariate or joint probability approach. In the univariate approach, data on the wave heights and water levels are transformed to the response function. This response function data then form the basis of a univariate extrapolation procedure. The joint

---

Box 4.3. continued

probability approach involves extrapolation of the joint probability density of the wave heights and water levels, before simulating many thousands of years worth of data (this is to avoid direct integration of the response function, which can be complex). These loading condition data are then transformed to the response function, and the distribution of the response is obtained.

Comparing the approaches

The loading condition approach is practical and simple to apply, however, the assumption that the return period of the loading conditions equals the return period of the response function can lead to under design, if the method is not well understood. The 200-year overtopping rate, which is obtained from the loading condition approach, will be significantly less than the 200-year overtopping from the response approach. This is because the loading approach does not consider all of the combinations of wave heights and water levels that can produce the 200-year response. The extent of the discrepancy varies with response function, the dependence level and return period. To account for the discrepancy, different measures can be applied and these include taking a conservative dependence value or simply adjusting the results by a 'typical' return period factor. The differences between the two approaches for overtopping at West Bay (Case study 6) are shown in the figure below.

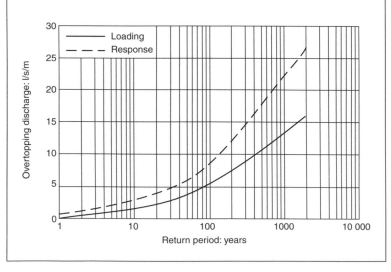

Table 4.3. *Current approaches to management of uncertainty in design*

Method to deal with uncertainty	Hydraulics		Geotechnics		Structures	
	Loading	Response	Loading	Response	Loading	Response
Sensitivity testing	✓*	✓	✓		✓	
Probabilistic modelling	✓✓†	✓✓	✓	✓	✓	✓
Lump sum factor of safety			✓✓	✓✓	✓	✓
Partial factor of safety			✓	✓	✓✓	✓✓

* Occasionally applied.
† Generally applied.

of a specific engineering solution. In terms of coastal and fluvial engineering, the common approach is moving from basing design assumptions on loading conditions to the response of the structure, or rather performance-based design, which is becoming more dominant particularly in North American design codes.

4.3. MOVING TOWARDS A COMMON RISK-BASED APPROACH

Previous design practices, as is observed in some of the case study analysis and through discussion with both clients and designers, are focused on breaking the problem down into manageable elements and using the appropriate discipline to design that element, then bolting the design together to provide a complete solution. However, the industry is now tending towards a systems approach to the design of coastal and fluvial structures that complements the philosophy of a common risk level by allowing differences between design disciplines and tools to be slowly smoothed out.

The range of structures that this book is attempting to satisfy means that a specific model for a common risk-based approach is difficult to prescribe. Also, a specific model is of little use to the majority of designers as each project will present its own set of challenges before even the interaction of the three main design disciplines is considered. Therefore, this chapter focuses on a generic design philosophy that needs to be considered in order to move towards a common risk level. The approach is based largely on the lessons learnt from the case study analysis (Part B) and through discussions with both experienced clients and designers.

The approach is split into four key stages (see Figure 4.1), as follows.

- *Design Stage 1: Define the structure's performance objectives and indicators*. The designer is focusing on setting out the parameters on which the design is going to be tested. An important aspect of this process is defining the design life of the structure, as the risk levels set for the other performance indicators will be based on this decision.
- *Design Stage 2: Design engineering system and define failure paths*. Set out the structural system by defining the individual structural elements. Then define the interaction between these

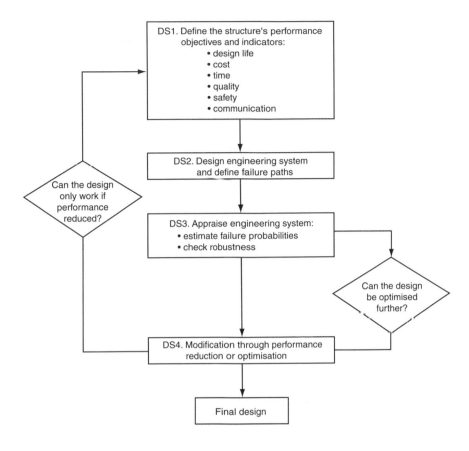

Figure 4.1. Detailed structural design process

elements in terms of leading to failure as defined by the performance indicators.

- *Design Stage 3: Appraise engineering system*. The designer needs to appraise the system by considering the probability of failure compared with the performance indicator and also to make checks on the robustness of the design in terms of spatial and temporal changes.
- *Design Stage 4: Modification through performance reduction or optimisation*. This last step allows the process to be iterative, with

the expectation that it will take a number of passes to optimise the design to an acceptable level by either modifying the performance objectives to a more realistic level or improving the structural system. This process focuses on the detailed design of the project cycle but transposes into the concept of the Acceptable Risk Bubble, with setting performance objectives being an essential part of the functional analysis stage.

The remainder of this chapter will discuss the key stages that the designer should consider during the detailed design and sets out the wider issues that should perhaps be considered by the designer in more detail at each stage. The main issues are:

- defining design life
- defining other performance objectives and indicators
- the relationship between design life, annual probability and lifetime probability
- the currently accepted performance indicators used in coastal and fluvial structures
- development of system diagrams, event trees and fault trees
- estimating failure probabilities of elements and total failure probabilities of systems
- checking the robustness of the structural system.

4.4. DESIGN STAGE IA: DEFINING A DESIGN LIFE

The definition of design life is the period of time the coastal or fluvial structure is expected to perform, given the design conditions, with normal maintenance. However, in practice design life is a complicated and sometimes confused term. Through discussions with clients and consultants, the following categories of design life are defined:

- service life — the period of time over which the client expects the structure to perform
- appraisal life — the period of time over which the client and respective funders or risk owners expect to see a return in their investment
- element life — the period of time over which a certain element will provide sufficient strength to the structure with or without maintenance.

The time period of these categories varies considerably depending on each project and should be assessed individually rather than prescribed from a standard value. The key issues that need to be considered are:

- the temporal need of the client and other risk owners (i.e. is the structure temporary or permanent?)
- the period in which the client/funder wishes to see a return on their investment
- the type of structure
- environmental conditions
- construction materials to be used.

However, whatever design life is decided upon it is critically important that the decision is clearly communicated to all parties, as decisions later on in the design process will rely largely on this decision. The key issues with each design life category are discussed in more detail below.

4.4.1. Service life

Guidance is available on the lifetime of certain structures in a number of codes of practice and other design manuals. The main ones that are relevant to hydraulic structures are as follows.

- In Eurocode 1 (BSI, 1996), the classification of the design working life is identified as a number of years against the type of structure (see Table 4.4).
- BS 6349: Part1 (BSI, 2000) on maritime structures recommends the following values as minimum requirements:
  - quay walls — 60 years
  - open jetties — 45 years
  - superstructure works — 30 years
  - dry docks — 45 years
  - shore protection works and breakwaters — 60 years
  - flood protection works — 100 years.
- The Highways Agency specify the service life of a bridge to be 120 years.

Table 4.4. Design working lives from the UK National Document for Eurocode 1 (BSI, 1996)

Class	Required design working life: years	Example
1	1–5	Temporary structures
2	25	Replaceable structural parts (gantry girders, bearings)
3	50	Building and other common structures other than those listed below
4	100	Monumental building and other special or important structures
5	120	Bridges

In a vast amount of cases, the true service life may effectively be considered indefinite. This is the case with many of the bridge crossings in London for example, where the disruption costs alone would be so severe that the thought of rebuilding the bridge again is not worth considering. Therefore, in these cases, the designer and client should select a service life, which is a reasonable indicator of the time period in which the structure will operate with only normal maintenance. In practice, of course, to aim for a service life greater than 120 years would generally be unrealistic because of either high construction costs associated with trying to build in the durability considered necessary or because of the uncertainty in the structure's role so far into the future.

4.4.2. Appraisal life

Appraisal life varies depending on the funders of the scheme, which can be categorised into two main categories as follows.

- *Public funding.* These are organisations such as local authorities and national government agencies, for example, the Environment Agency and DEFRA. The appraisal life is generally set over a long period to a limit where discounted annual damages have limited impact on the BCR. In general, the appraisal life for publicly funded schemes such as flood protection is defined as 50 years.

- *Private funding*. These are organisations, such as port authorities and developers, who are looking to generate money through the construction of a scheme and therefore require return on their investment over a much shorter time-scale.

A common mistake made by designers is to equate appraisal life with service life, which in some cases may be identical. However, the two are quite separate concepts and should be both defined clearly by the designer early on in the project cycle through negotiation with the client (see Chapter 3).

4.4.3. Element life

The lifetime of an element is related directly to the material selected, the level of exposure of the material in-situ and its durability in that structural form (discussed in Section 4.2). The designer, therefore, has two design choices when setting the element life:

- ensure the durability of the element will provide sufficient strength until the end of the service life
- install an inspection and maintenance regime (with replacement as necessary) so that the required level of strength is maintained throughout the service life.

To obtain typical life spans of elements in coastal and fluvial structures, the reader should contact manufacturers and consult records of existing structures. It is clearly a good idea for clients and designers to build up their own knowledge base of element lifetimes. Local engineering knowledge should always be used wherever possible to inform judgements.

4.5. DESIGN STAGE IB: DEFINING OTHER PERFORMANCE OBJECTIVES AND INDICATORS

4.5.1. Defining other performance objectives

In Chapter 3, we discussed the concept of the scheme or system having multiple attributes that need to be balanced. To achieve that balance, a number of conceptual designs will be produced in order to select the most appropriate to be taken forward to detailed structural design. With the most appropriate design selected, the designer can focus then on the main structural performance objectives as an outcome in terms of:

- cost (e.g. £/annum)
- time (e.g. number of hours per year full service available)
- quality (e.g. overtopping — 'free' days per year)
- safety (e.g. days free of accidents during construction, maintenance and operations).

As an example, the main performance objectives in terms of structural response of the three main case studies analysed were:

- Case study 1, Clacton sea defences — no flooding through overtopping or breaching
- Case study 2, Castle Cove — no major landslides causing loss of property
- Case study 3, Dawlish — no delay to the operation of the railway.

The performance objectives must be achievable and map onto the Acceptable Risk Bubble for the particular project. The aim of the Acceptable Risk Bubble is to aid in the decision-making process of selecting a design, while this process is focused on setting out a common risk-based design to optimise structural performance. Therefore, some of the softer issues in the acceptable bubble may not be transposed down to this level, such as improvements to tourism, as they can be managed more effectively at the project level.

4.5.2. Defining performance indicators

In many ways it would be better to exclude probability from performance and state that performance relates to achievement. However, in practice the structural performance indicator needs to be subdivided into an expectation and the probability of reaching that expectation. The level of expectation can be subdivided into a range of levels from serviceability issues such as access, general operation of the structure through to its ultimate limit (i.e. complete failure).

For example, a designer of a flood defence may have a range of performance indicators under the same objective of wave-induced overtopping. The indicators may be:

- for the 1 in 1 year event, wave-induced overtopping must allow safe access to crest
- for the 1 in 200 years event, no significant flooding should occur
- for the 1 in 2000 years event, significant flooding may occur but structural damage will be minimal.

This example shows that the choice regarding the level of performance is a subjective choice and the choice made will depend on the particular personal situation of the designer and the client (i.e. a flood defence engineer may use different levels compared to a port engineer). In this chapter, we present comparisons of accepted risk levels for different situations that illustrate the importance of risk perception further.

There are a number of documents that provide indicative and target standards to guide designers, and it is important that the designer understands the fundamentals behind these documents in order to make appropriate decisions with regard to setting risk levels. The main aspects that the designer needs to understand are:

- the relationship between design life, annual probability and lifetime probability
- the accepted performance indicators currently used in coastal and fluvial structures.

4.5.3. The relationship between design life, annual frequency of exceedance and lifetime probability

A survey undertaken for a framework study on *Common risk based design standards* funded by the Department of the Environment (Piechowiak and Simm, 1997) found that the concept of design life and its relationship to probability was the most confused and misunderstood in the design process. This concept needs to be clearly explained and understood before the designer can start to consider having any influence on setting appropriate risk levels.

In coastal and fluvial design, normal practice is to consider exceedance probabilities, defined as the probability of an event being exceeded over a specific time period. These time periods can vary but the most common ones are annual and lifetime. It is worth noting that the inverse of an annual probability transposes into a return period measured in years. The annual return period is currently the most common measure of performance (e.g. flood protection set to resist the 1 in 200 year event). However, return period to a lay person can be misunderstood because some believe that, say, a 1 in 200 year event once happened will not occur again for another 200 years. The reality is that the event is only a statistical representation of exceedance and as a consequence the same event or worse could happen the very next day. To mitigate against this confusion, a move towards the use of lifetime probability is starting to occur, especially in more recent UK Government guidance documents (see Planning Policy Guidance (PPG) 25 (DoE and Welsh Office, 2001)). The main benefit is that the lifetime probability encapsulates both design life and annual return period, so that a wide range of schemes can be compared using the same measure. The lifetime probability of exceedance can be calculated using the following formula:

$$P = 1 - \left(\frac{1-1}{TR}\right)^n$$

where TR is the design return period and n is the design life.

This formula has been calculated for a number of different combinations of design return period and design life, as shown in Table 4.5.

The main confusion found in the previous framework study (Piechowiak and Simm, 1997) was that some designers were setting the design return period to the equivalent design life. Table 4.5 shows that the probability of a 50-year return period being exceeded during a 50-year design life is 0·63 which, if selected as the design event, still has a reasonable chance to occur. Although this may no longer be a common error by many designers, some still manage to make this error, leaving their design open to under-performance. It is also worth noting that, over time, a number of failures of hydraulic structures have occurred because too low a design event was selected. Therefore, the designer needs to understand clearly the relationship between design life and probability of both annual and lifetime failure.

An example of this point is described in the Tekirdag Port case study (Case study 5) where the design life and the design return period were the same. In this case, the design was sufficiently robust (i.e. adequate factor of safety) to withstand more extreme events that

Table 4.5.   The probability of a particular return period event being exceeded during the design life

Design return period: years	Design life: years						
	2	10	50	100	200	500	1000
2	0·63	0·99	1·00	1·00	1·00	1·00	1·00
10	0·18	0·63	0·89	1·00	1·00	1·00	1·00
50	0·04	0·18	0·63	0·86	0·98	1·00	1·00
100	0·02	0·10	0·39	0·63	0·86	0·99	1·00
200	0·01	0·05	0·22	0·39	0·63	0·92	0·99
500	0·00	0·02	0·10	0·18	0·33	0·63	0·86
1000	0·00	0·01	0·05	0·10	0·18	0·39	0·63

could be experienced at the site. In the Sea of Marmara in Turkey, wave fetches are limited and tidal variations small. However, if the same design had been undertaken on the south-west coast of the UK then the influence of wave height and water level would have been more significant because of the greater tidal range, including storm surges, and larger waves from the Atlantic. In fact, for Case study 5, the error in selecting a frequently occurring design condition had little impact on the performance of the structure (Figure 4.2). However, as the scenario testing showed, this might not have been the case for all structures in all situations so care should be taken by the designer in selecting an appropriate risk level.

4.5.4.   The currently accepted performance indicators used in coastal and fluvial structures

A vast amount of research has been undertaken in defining indicative standards and target probabilities of failure for the design of structures, especially for buildings where minimising the chance of failure in terms of human life has been a priority over the last 50

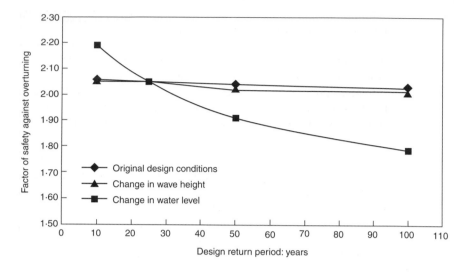

Figure 4.2. Impact on the factor of safety with events above the accepted design (Case study 5)

years. Box 4.4 shows that under the Eurocode 1 (BSI, 1994), the expected lifetime probability for ultimate failure is between 10^{-3} and 10^{-7}, which is a reasonable reflection of the standards used currently in structural engineering.

The UK Government specifies guidance on the indicative standards used in coastal and fluvial structures, which is generally the most applicable to the subject of this book. The indicative standards of protection specified in England and Wales are based on economic performance of flood defence (see Box 4.5). The UK Government also state that there should be 'no loss of life' but, in the economic calculations, no quantification in monetary terms of societal risk is made, therefore, the indicative lifetime probability of failure is much lower than those quoted in Eurocode 1 (see Box 4.6). However, this figure is slightly misleading because the risk of fatality is managed through flood warning and evacuation rather than limiting the structural response. In certain locations or other countries, evacuation may not be a viable strategy and therefore Eurocode 1 may be a more appropriate level to work to. Eurocode 1 is most definitely more

Box 4.4.   Eurocode 1: Part 1 (BSI, 1996)

In Eurocode 1, Part 1 recommends total probability of failure over design working life for both ultimate and serviceability limit states that also consider the consequence of failure. The fundamental requirements in Eurocode 1 are that the given properties of the structure will be suited for its intended use. These requirements are specified as such:

- with respect to serviceability, a structure shall perform its required function during an agreed percentage of time and remain fit to do so during its planned lifetime
- with respect to ultimate strength, a structure shall sustain all actions and influences likely to occur during execution and use
- with respect to robustness, a structure shall not be damaged to an extent that is disportionate to the original cause.

Consequence of failure is classified as:

- minor — risk to life, given a failure, is low and also economic and social consequences are small or negligible
- moderate — risk to life, given a failure, is medium or economic and social consequences are considerable
- large — risk to life, given a failure, is high, or economic and social consequences are very great.

The code mentions that a fourth class could be added for extreme consequences (e.g. storm surge barriers or nuclear power points). For each failure category, the code defines the target lifetime probability of failure for three relative costs of safety measures or structural strengthening (i.e. the cost of strengthening against the benefit). The three levels are:

- high — high costs against low benefits
- moderate — reasonable costs against reasonable benefits
- low— low costs against high benefits.

Combining all of these aspects, Eurocode provides target lifetime probabilities for a range of safety measure and limit states. The code suggests that a high cost/low benefit structure can be designed to a higher risk level compared to a low cost/high benefit structure (see table overleaf).

Box 4.4. continued

Target probability of failure for design working life as defined by Eurocode 1 (BSI, 1996)

Relative cost of safety measures	Serviceability limit state	Ultimate limit states		
		Expected consequences of failure		
		Minor	Moderate	Large
High	0·160	$2·6 \times 10^{-3}$	$4·8 \times 10^{-4}$	$7·2 \times 10^{-5}$
Moderate	0·067	$4·8 \times 10^{-4}$	$7·2 \times 10^{-5}$	$8·5 \times 10^{-6}$
Low	0·023	$7·2 \times 10^{-5}$	$8·5 \times 10^{-6}$	$7·9 \times 10^{-7}$

suited to traditional structural design, as safe evacuation of a building before it collapsed would be more difficult to manage than that of properties in a flood zone as there would be some warning prior to the event (i.e. a weather forecast).

The reality observed in the case study analysis (Part B) showed that the actual accepted lifetime probability of occurrence of the design condition ranged from certainty to as low as 0·05 for individual structural elements (Table 4.6). If the overall system was considered (i.e. the chain of events leading to failure) then the range of lifetime failure probabilities reduced to between 0·63 and 0·02 (Table 4.6). Further details on this back analysis are presented in Case studies 1–3 in Part B. It is worth noting that the case studies ranged from moderate costs/moderate benefits to low costs/high benefits. It must be stressed that this analysis is simplistic and was undertaken to demonstrate a systems-based approach to design — new designs must always be undertaken using the most appropriate techniques.

Three aspects can be concluded from this overall analysis. First, both best guidance and the case studies generally focus on setting risk levels in terms of resistance to a loading condition. However, the designer is reminded to consider one of the issues highlighted in

Box 4.5. Indicative standards for flood protection in the UK (MAFF, 1999)

In the UK, DEFRA is responsible for funding flood and coastal protection with the operational elements of flood defence being delegated to the Environment Agency and coast protection being delegated to local authorities. As part of their responsibility, DEFRA have set indicative standards for flood and coast protection using the following criteria:

- fluvial or coastal source
- economic worth of the area protected.

The indicative standards are published in the *Flood and coastal defence project appraisal guidance — economic appraisal* or FCDPAG3 (MAFF, 1999), but for the benefit of the reader the two main tables have been reproduced below and overleaf. A simple risk calculation is described in Box 4.5 showing the variation in relative risk level between land use bands. Also, a comparison between DEFRA guidance and Eurocode 1 is made.

Indicative standards of protection defined by FCDPAG3 (probabilities derived using Table 4.5)

Land use band	Indicative standards of protection			
	Fluvial		Coastal/saline	
	Return period: years	Lifetime probability of failure*	Return period: years	Lifetime probability of failure*
A	50–200	0·64–0·22	100–300	0·39–0·15
B	25–100	0·87–0·39	50–200	0·64–0·22
C	5–50	1·00–0·64	10–100	1·00–0·39
D	1·25–10	1·00	2·5–20	1·00–0·92
E	<2·5	1·00	<5	1·00

* Based on a 50-year design life (i.e. design life assumed equal to appraisal period)

Box 4.5.   continued

Description of land use bands defined by DEFRA (MAFF, 1999)

Land use band	Indicative range of housing units (or equivalent) per km of coastline or single river bank	Description
A	>50	Typically intensively developed urban areas at risk from flooding and/or erosion
B	>25–<50	Typically less intensive urban areas with some high-grade land and/or environmental assets of international importance requiring protection
C	>5–<25	Typically large areas of high-grade agricultural land and/or environmental assets of national significance requiring protection with some properties also at risk, including caravans and temporary structures
D	>1·25–<5	Typically mixed agricultural land with occasional, often agriculturally related, properties at risk. Agricultural land may be prone to flooding, water logging or coastal erosion. May also apply to environmental assets of local significance
E	>0–<1·25	Typically low-grade agricultural land, often grass, at risk from flooding, impeded land drainage or coastal erosion, with isolated agricultural or seasonally occupied properties at risk, or environmental assets at little risk from frequent inundation

Box 4.3, namely that accepting a loading condition risk may involve accepting a much higher risk level in terms of structural response.

The second aspect is that the distribution of probabilities for individual structural elements can have a high variance. However, when the element is considered as a part of a system then the high risk element relies on other elements to provide adequate protection (i.e. a number of things have to go wrong before that weak element fails itself). In coastal and fluvial engineering, the designer must make clear choices regarding the management of risk for a particular element. The choice may be to build more strength into that element or to protect it in some way. As design of coastal and fluvial structures moves to a more systems-based approach, there may be further opportunity to reduce whole-life costs by optimising structural elements so that the overall risk level of the system is not compromised. In some cases, for example the refurbishment of a seawall, where it many be more cost-effective to invest in one element rather than invest in raising all of the structural elements to the same risk level, this illustrates that the designer needs to focus on the structural system rather than trying to raise or lower risk levels of a particular element.

Box 4.6. Differences between relative risk levels using DEFRA guidance (MAFF, 1999) and Eurocode 1

The simple calculation in the table overleaf shows that under the current grant aid guidelines set out in FCDPAG3 (MAFF, 1999) there is an implied attempt by the UK Government to set a common risk level for flood protection. The approach involves increasing the tolerable probability of failure as economic consequences reduce. The calculation of the resulting risk value remains stable (at approximately £30 000) for the high value land uses (A to C). Having said this, inconsistency starts to creep in for the lower value land use bands (D to E), where the relative risk is permitted to reduce.

Flooding from overtopping or other non-structural failures can be defined as a serviceability limit state because the structure remains intact and continues to operate after the overtopping event. The following comparisons between the DEFRA guidelines and Eurocode 1 can be made (see table overleaf).

Box 4.6.   continued

- Band A can be defined as having a *low* relative cost of safety measures (low cost/high benefit), which under Eurocode 1 sets a target probability of failure at 0·023. Combining this value with the indicative damage value of £160 850 provides a relative risk of £3699, compared with £35 658 for the DEFRA guidelines, where the target failure probability is 0·22.
- Band D can be defined as having a *high* relative cost of safety measures (high cost/low benefit), which under Eurocode 1 sets a target probability of failure at 0·16. Combining this value with the indicative damage value of £9651 provides a relative risk of £1544, compared to £9559 for the DEFRA guidelines, where the target failure probability is 0·99.

This comparison with DEFRA guidelines shows that the acceptable relative risk is lower by a magnitude of approximately ten if target probabilities are used as defined by Eurocode 1. However, to say that risk levels accepted in the UK are not compliant with Eurocode 1 and are therefore not safe (if we assume that Eurocode 1 provides an adequate level of safety) is perhaps too strong a point of view. It must be noted that Eurocode 1 is defined for buildings rather than coastal and fluvial structures where the consequences of a building collapsing is perceived to have a larger consequence (i.e. individuals may expect a flood defence to fail but they do not expect their house to fall down).

The acceptable risk level is not the same in other European countries. For example, in the Netherlands the optimal probability of failure over a 50-year working design life is 0·005 (10 000 years return period) for Band A areas, which means that a lower risk level than the Eurocode 1 serviceability limit state is required. The main reason for this lower risk level is perhaps related to difficulties in the Netherlands of reducing flood risk by building outside the flood risk zone or allowing areas to be quickly evacuated. In the UK, these aspects can be managed through appropriate planning control (i.e. reduce building on the floodplain) and with safe areas generally near to risk areas, and flood forecasting can be used to make sure areas are evacuated when required. The designer is advised to consider local advice on the subject and look at a wide range of risk management strategies before focusing on a complete structural solution.

Box. 4.6.   continued

Calculation of relative risk for coastal flooding using DEFRA guidelines

Land use band	Average no. of housing units	Indicative flood damage if flooding occurs per km of defence: £*	DEFRA guidelines		Eurocode 1 guidelines	
			Probability of occurrence during a 50-year working life	Relative risk: £	Probability of occurrence during a 50-year working life: £	Relative risk: £
A	50	160 850	0·22	35 658	0·023	3700
B	35	112 595	0·33	37 242	0·069	7732
C	15	48 255	0·60	28 975	0·114	5517
D	3	9651	0·99	9559	0·160	1544
E	1	3217	1·00	3217	0·194	625

* The risk calculation assumes that the indicative flood damage is based on a housing equivalent of £3217 (Environment Agency, Southern Region).

Table 4.6.   Summary of the case study back analysis showing accepted design levels

Case study (CS)	Design life: years	Estimated design return period: years	Lifetime probability of failure for an individual element*	Conditional lifetime probability of failure if element is considered part of system†
Clacton (CS1)				
Beach control structure	50	150	$0.28\ (P_1)$	$0.28\ (P_1)$
Beach	10	150	$0.06\ (P_2)$	$0.02\ (P_1 \times P_2)$
Seawall	50	15	$0.96\ (P_3)$	$0.02\ (P_1 \times P_2 \times P_3)$
Castle Cove (CS2)				
Rock revetment	60	100	$0.45\ (P_1)$	$0.45\ (P_1)$
Coastal slope	60	1500	$0.04\ (P_2)$	$0.02\ (P_1 \times P_2)$
Dawlish (CS3)				
Beach	50	1000	$0.05\ (P_1)$	$0.05\ (P_1)$
Seawall structure	50	200	$0.22\ (P_2)$	$0.01\ (P_1 \times P_2)$
Seawall crest	50	7	$0.99\ (P_3)$	$0.01\ (P_1 \times P_2 \times P_3)$
Track support	3	1	$1.00\ (P_4)$	$0.01\ (P_1 \times P_2 \times P_4)$
Coastal slope	3	1	$1.00\ (P_5)$	$0.01\ (P_1 \times P_2 \times P_5)$
Turkey (CS5)				
Quay wall stability	50	50	N/A	0.63

$*P_{individual} = 1 - \left(1 - 1/TR\right)^n$ where TR is the design return period and n is the design life.

$†P_{total} = P_1 \times P_2$ and so on, assuming that the failure of each element is conditional.

The final aspect is that there are clear indications of an indicative lifetime probability of failure for a system, which is widely acceptable within the coastal and fluvial engineering community. The lifetime probability accepted is inversely proportionate to the consequence, i.e. as the consequence reduces then the lifetime probability will increase until failure is certain and the only viable option remaining is 'do nothing'. However, at the other end of the spectrum, the back analysis of the case studies suggest that there may be two categories of tolerability, high and low, that the designer can apply to the design of their systems. Tolerability is defined as the amount of time or resource available to amend the problem when failure occurs. High tolerability is where sufficient time is available to fix the problem or where a failure of an individual component does not form an immediate catastrophic failure to the system (e.g. a rock structure could suffer movement and can be repaired). Low tolerability is where a rapid catastrophic failure of the system occurs and there is little time to minimise the consequence of failure (e.g. a beach where the designer was trying to maintain a consistent standard throughout the design life). Using this concept, target lifetime probabilities can be suggested for coastal and fluvial structures that have moderate costs/moderate benefits (Table 4.7).

The key message is that the designer must consider a wide range of issues when selecting the risk levels for their structure, including risk management strategies, interaction of structural elements and the tolerance of failure.

Table 4.7. *Indicative probability of failure during the design life from back analysis*

Level of tolerability	Possible examples	General description of the limit state	Indicative lifetime probability of occurrence
High	Rock revetment Beach with maintenance	Structure starts to fail but will take time to reach a catastrophic state allowing time to make repairs or undertake another contingency plan	0·1–0·2
Low	Concrete material Coastal landslide Beach without maintenance	Rapid catastrophic failure of the structure allowing no time for minor repair	0·05 or less

4.6.   DESIGN STAGE 2: DESIGN ENGINEERING SYSTEM AND DEFINE FAILURE PATHS

Once the designer has set out the performance objectives and indicators for the option, the next stage is to design the structural system that will satisfy those parameters. As discussed previously in Chapter 3, systems showing the fault paths and interaction of risks can be mapped or sketched through the development of fault and event trees. These aid the designers in their understanding of the problem. Failures in design have generally come from a poor understanding of how the structural elements and the associated risks interact. This failure has resulted in the discounting of some risks as being too improbable or the consequence not significant but in fact they may be the start of a chain of events that lead to a large catastrophic failure. A systems based approach allows the designer to focus on the key risks within the design in order to meet the performance objectives.

A classic example is the refurbishment of a seawall where the masonry needed repair and the decision was made to re-point the structure to increase its structural integrity. Some time later the seawall overturned because of the inadequate drainage through the wall due to all the gaps being grouted up, which allowed excessive pore water pressures to build up sufficiently to force failure. Here the designer had focused on the risk of collapse through one specific risk — structural strength — rather than looking at all the other possible risks to the wall and creating a solution that managed all of the risks appropriately.

In the case study analysis (Part B), there are a number of system diagrams, event trees and fault trees drawn to illustrate the structural system adopted. One example from the case study analysis presented in Part B is the system diagram developed for the Clacton sea defences (Case study 1) that shows the interaction between the beach, groyne and seawall (repeated here as Figure 4.3)

4.7.   DESIGN STAGE 3A: ESTIMATE FAILURE PROBABILITIES OF ELEMENTS AND TOTAL FAILURE PROBABILITIES OF SYSTEM

As discussed in Section 4.5.4, there is now a move from using the terminology of annual return periods to lifetime probabilities. Estimating these failure probabilities may use a mixture of design

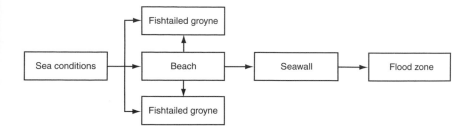

Figure 4.3. System diagram of the Clacton sea defence (Case study 1)

tools, as described in Table 4.1. This is where management of uncertainty is very important so that the probabilities estimated are realistic. This process will always require the use of experienced engineers to ensure that the assumptions and methods used are encapsulating best practice at the time. An example of estimating probabilities can be found in the Castle Cove coast protection case study (Case study 2). The case study demonstrates the principals behind estimating the probabilities of each structural element failing in order that the total annual probability of failure could be estimated (Figure 4.4).

4.8.   DESIGN STAGE 3B: CHECKING THE ROBUSTNESS OF THE ENGINEERING SYSTEM

This is perhaps the most important part of the detailed design process as the success of most coastal and fluvial designs are their durability through time and their performance spatially.

4.8.1.   Understanding the temporal nature of risk

Designers may feel that if a system works on day one or has performed satisfactorily for many years then it must be safe. In practice, risks may increase due to fatigue, increased loading (potentially through climate change) and inadequate maintenance. The designer needs to understand how the risks involved with individual elements may change over time allowing an appropriate management strategy to be organised. This may lead to a decision to either strengthen that element

$P_{total} = P(I) + P(II)$

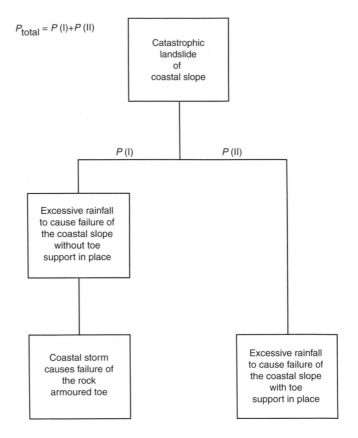

Figure 4.4. Castle Cove fault tree with estimation of probabilities

so the risk at the end of its design life is still acceptable, or to consider a programme of maintenance or replacement instead to maintain the acceptable risk level (see Case study 1). With the latter, the designer needs to consider how this maintenance will be undertaken not only in terms of the engineering challenges, but also in terms of structural and personal safety (Joyce, 2001).

4.8.2. Understanding the spatial nature of risk

In Chapter 3, the importance was stressed of defining the correct physical boundaries so that spatially the correct length or area is considered in the design process, ensuring that a common risk level is maintained. Once the physical boundary has been established, it is important for the designer to understand how specific risks may change spatially.

For example, with the design of seawalls (or any other linear structure), it is common design practice to select a nominal cross-section for a particular length or sub-length of seawall. The designer needs to ensure that:

• the design sub-lengths are defined such that there is sufficient difference in risk level between sub-lengths to justify a difference in design, with the associated additional cost or cost-saving compared with the previous section
• for any particular sub-length of structure, the cross-section with the highest risk is selected

There have been situations where designers have failed to appreciate the importance of spatial risk, which in the extreme may result in:

• failure due to the selection of the wrong cross-section locations or loadings
• designs that have a significant amount of redundancy in them.

Certainly, weaknesses in the designed structure should be discouraged but it is better to be aware of them and therefore take appropriate measures to remove or mitigate the risk. High redundancy is perhaps acceptable but a balance between construction costs and design costs, or the increased cost in construction due to more cross-section changes, should be considered. It is, therefore, good design practice to ensure that the limits of spatial risk are within acceptable bands (see Case study 4 for an example).

4.9. DESIGN STAGE 4: MODIFICATION THROUGH PERFORMANCE REDUCTION OR OPTIMISATION OF THE STRUCTURAL SYSTEM

Designing is an iterative process and through the design process the designer may be faced with a design that does not meet the performance criteria or alternatively a design with scope for optimisation. The latter is a matter of passing through the process again, while the former requires some thought regarding the reasons why a performance objective should be changed.

The primary reason for the designer to change the performance indicators is where the structural system is unable to perform to the standard prescribed. Before changing these indicators, the designer must question whether safety is being compromised and that another system should be sought. Also the designer may need to return to the stakeholders to make changes if they conflict with the agreements made during the establishment of the Acceptable Risk Bubble. In some cases there may also be a clear argument to raise standards. So the designer has two main approaches that can be followed.

- *Modify design life.* As discussed previously, the design life is intrinsically linked to the lifetime probability of failure of the system. If the designer wishes to keep the lifetime probability of failure constant, but is unable to meet performance, then one solution is to reduce the design life.

  A simple example of this is demonstrated in the bridge scour case study (Case study 7) where the acceptable risk level can be influenced by varying the design life of the bridge (Figure 4.5). However, in this particular case study, the structure's design life was 120 years and the design condition for scour protection was the hydraulic event with a 200-year return period. However, the optimum risk level for 120-year design life is a hydraulic event with an 800-year return period and, hence, more investment in scour protection could potentially reduce the whole-life cost of the bridge.

- *Modify the performance objectives and accept the additional risk.* The other alternative is to change the performance objectives and accept any additional risk of doing so. In some cases, this may be quite a rational choice in order to make a design work. As mentioned above, care needs to be taken to ensure that the safety levels are not pushed to an unacceptably dangerous position. One example of where the client made an informed choice about reducing the safety levels in order to reduce costs and aesthetic impacts on the area is presented in Case study 8.

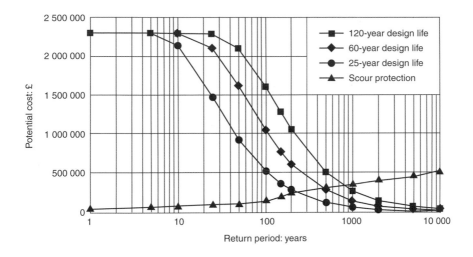

Figure 4.5. Risk of failure against the capital cost of scour protection (Case study 7)

4.10. SUMMARY: KEY ASPECTS OF APPLYING THE RISK FRAMEWORK ON A SPECIFIC ENGINEERING SOLUTION

The specific uncertainties in coastal and fluvial engineering may be categorised in relation to the design tools used, variation in hydraulic conditions, variation in ground conditions and reliability of materials' strength and durability. These issues were discussed earlier, and this chapter then considered the main emphasis in the three main disciplines (hydraulics, geotechnics and structures) and how each one manages the specific uncertainties.

The main differences between the design disciplines are:

- structures — heavily codified, the use of partial safety factors
- geotechnics — often empirical design codes, limited use of partial safety factors
- hydraulics — limited codes, largely dependent on guidance notes.

In terms of setting design risk levels, it is incredibly difficult to generalise because of the uncertainties and complexities of the design tools and disciplines involved. The main focus of this chapter has been to set out the key steps that the designer should follow to make robust decisions with respect to setting appropriate risk levels. The key stages as set out in Figure 4.1 are as follows.

- *Stage 1: Define structural performance objectives and indicators*. The designer needs to be clear in the definition of the design life and failure probabilities. The recommendation of this book is to use the terminology of lifetime probabilities rather than annual return periods. The latter is misleading to the lay person about the risk level being accepted. The designer should make sure failure probability terminology and numbers used are clearly understood and communicated to the stakeholders of the project.

  An effort has been made in this book to set out an indicative lifetime probability of failure for moderate costs/moderate benefits projects or above. The values are split into two levels of tolerability — high (can afford some failure to occur) or low (any failure will quickly lead to catastrophic failure). The indicative lifetime probabilities are set at 0·1–0·2 and less than 0·05, respectively.

  The designer must also consider if performance is to be based on loading or structural response. The majority of design methods are tailored to design loading, but as design tends to a more systems based approach, tailoring to structural response may be a more suitable approach. If loading is used then the risk levels should remain higher than those used for structural response due to the uncertainty of the system's overall performance.

- *Stage 2: Design structural system and define failure paths*. Sketching the system helps the designer to select the appropriate structural elements required in the system and help manage the risk management process.

- *Stage 3: Appraise structural system*. The designer needs to consider how risk levels may change temporally and spatially, such that the performance of the structure remains within acceptable bands.

- *Stage 4: Modification through performance reduction or optimisation of the structural system.* The designer has the flexibility to make changes by either modifying the performance objectives, such as design life, to make the structural system acceptable or by optimising the existing system to fit within the performance limits.

References

5

5. References

Bowland B. J. and Derghin J. C. (1998). *Robust estimates of value of statistical life for developing economies — an application to pollution and mortality in Santiago*. Department of Economics, Iowa State University.Technical Report.

British Standards Institution (BSI) (1999). *Codes of practice for site investigations*. BSI, London. BS 5930.

British Standards Institution (BSI) (1995). *BS ENV 1997 Eurocode 7: Geotechnical design — Part 1. General rules*. BSI, London

British Standards Institution (BSI) (1996). *BS ENV 1991 1 Eurocode 1: Basis of design and action on structures — Part 1. Basis of design*. BSI, London.

British Standards Institution (BSI) (1997). *Code of practice for structural use of concrete — Part 1. Design and construction*. BSI, London. BS8110.

British Standards Institution (BSI) (2000). *Code of practice for maritime structures — Part 1.General criteria*. BSI, London. BS6349-1

Bucciarelli, L. L. (1994). *Designing engineers*. MIT Press, Cambridge, Massachusetts.

Bye, P. (1998). *Easter 1998 Floods, Report by the Independent Review Team to the Board of the Environment Agency*. Environment Agency, Bristol.

Construction Industry Research and Information Association (CIRIA) (1977). *Rationalisation of safety and serviceability factors in structural codes*. CIRIA, London. CIRIA Report 63.

Construction Industry Research and Information Association (CIRIA) (1996). *Beach management manual*. CIRIA, London. CIRIA Report 153

Construction Industry Research and Information Association (CIRIA) (2000). *C561CD — RiskCom. Software tool for managing and communicating risks*. CIRIA, London.

Department of the Environment and the Welsh Office (2000). *Planning Policy Guidance: Development and flood risk, PPG25.* Department of the Environment and the Welsh Office, London.

French, S. (1988). *Decision theory: an introduction to the mathematics of rationality.* Ellis Horwood, Chichester.

Hammit J. K., Liu J.-T. and Lui J.-L. (2000). Survival is a luxury good: the increasing value of statistical life. *NBER Summer Institute Workshop on Public and the Environment*, NBER, Cambridge, Massachusetts.

Health and Safety Executive (HSE) (1999). *Reducing risks, protecting people.* HSE, London.

Heinrichs, P. and Fell, R (1994). Acceptable risks for major infrastructure. In Henrichs, P. and Fell, R. (eds) *Proceedings of the seminar on acceptable risks for extreme events in planning and design of major infrastructure*, Sydney, Australia. Balkema, Rotterdam.

Joyce, R. (2001). *The CDM Regulations explained*, second edition. Thomas Telford Publishing, London.

Meadowcroft, I. C., Hall, J. W. and Ramsbottom, D. M. (1997). *Application of risk methods in flood and coastal defence — a scoping study.* HR Wallingford, Wallingford. HR Report SR 483.

Ministry of Agriculture, Fisheries and Food (MAFF) (1999). *Flood and coastal defence project appraisal guidance — economic appraisal* (FCDPAG3). MAFF, London.

Ministry of Agriculture, Fisheries and Food (MAFF) (1999). *Flood and coastal defence project appraisal guidance — environmental appraisal* (FCDPAG5). MAFF, London.

Ministry of Agriculture, Fisheries and Food (MAFF) (2000). *Flood and coastal defence project appraisal guidance — approaches to risk* (FCDPAG4). MAFF, London.

Morris, M. W and Simm, J. D. (2000). *Construction risk in river and estuary engineering — a guidance manual.* Thomas Telford, London (for HR Wallingford and the DETR).

Nathwani, J. S., Lind, N. C. and Pandey, M. D. (1997). *Affordable safety by choice: the life quality method.* Institute for Risk Research, University of Waterloo, Waterloo, Ontario.

Nicholls, D. B. (1997). A framework for setting risk criteria in aviation, Advances in Safety and Reliability, Volume 2. *Proceedings of ESREL '97*, Lisbon. Pergamon, pp. 915–922.

Piechowiak, A. and Simm, J. (1997). *Common risk based design standards.* HR Wallingford (for the Department of the Environment), Wallingford. HR Report SR 504.

Simm, J. D and Cruickshank, I. C. (1998). *Construction risk in coastal engineering.* Thomas Telford Publishing, London (for HR Wallingford and the DETR).

Thomas, R. S. and Hall, B. (1992). *Seawall design.* Butterworth Heinemann (for CIRIA), London.

Vrijling, J. K. and van der Gelder P. H. A. J. M. (1999). *Uncertainty analysis of water levels on Lake Ijssel in the Netherlands: A decision making analysis.* Society for Risk Analysis, 9th Annual Conference — Risk Analysis: Facing the New Millenium. Society for Risk Analysis, Rotterdam.

Bibliography

6

6. Bibliography

American Society of Civil Engineers (1992). *Proceedings of the Short Course on Design and Reliability of Coastal Structures, 23rd International Conference on Coastal Engineering*, Venice, 1–3 October. America Society of Civil Engineers.

Birkenshaw, M. (1994). *Major accidents in the civil and structural engineering fields with specific reference to offshore.* Diploma in Occupational Health and Safety Management, Loughborough University of Technology, Loughborough.

Birkenshaw, M. (1999). *Presentation notes on setting of target safety levels.* Health and Safety Executive, London.

Blockley, D. (1985). *Reliability or responsibility?* Department of Civil Engineering, University of Bristol.

Bradford, P., McNair, I. and McNulty, A. (1999). A regulator's viewpoint — regulatory approach to judging the adequacy of nuclear safety-related civil engineering design. *Nuclear Energy*, **38**, No. 1, Feb., 41–46.

British Standards Institution (BSI) (1996). *General principles on reliability for structures — revision of first edition.* BSI, London. ISO 2394.

Chartered Institution of Water and Environmental Management (CIWEM) (1999). *Flood forecasting, warning and response.* Rivers and Coastal Group, Winter Meeting, January.

Chicken, J. C. and Posner, T. (1998). *The philosophy of risk.* Thomas Telford Publishing, London.

Construction Industry Research and Information Association (CIRIA) (1990). *Approach to risk assessment.* CIRIA, London. CIRIA Report 568.

Construction Industry Research and Information Association (CIRIA)/ Netherlands Centre for Civil Engineering Research Codes (CUR) (1991).

Manual on the use of rock in coastal and shoreline engineering. CIRIA, London. CIRIA Special Report 83/CUR Report 154.

Construction Industry Research and Information Association (CIRIA) (1999). *A guide to risk management of UK reservoirs*. CIRIA, London.

Cruickshank, I. C. and Hall, J. H. (2000). *Developing a risk communication tool (RiskCom)*. CIRIA, London. CIRIA Funders Report /CP/89.

Delft University of Technology (1997). Collection of publications. Delft University of Technology, Hydraulic and Offshore Engineering Section.

Delft University of Technology (1998). Collection of publications. Delft University of Technology, Hydraulic and Offshore Engineering Section.

Department of the Environment and the Welsh Office (1990). *Planning Policy Guidance: Development on unstable land, PPG 14*. Department of the Environment and the Welsh Office, London.

Department of the Environment and the Welsh Office (1992). *Planning Policy Guidance: Coastal Planning, PPG20*. Department of the Environment and the Welsh Office, London.

Department of the Environment and the Welsh Office (1994). *Planning Policy Guidance: Nature conservation, PPG09*. Department of the Environment and the Welsh Office, London.

Department of the Environment and the Welsh Office (1994). *Planning Policy Guidance: Planning and the historic environment, PPG15*. Department of the Environment and the Welsh Office, London.

Department of the Environment and the Welsh Office (1996). *Planning Policy Guidance: Development on unstable land: Landslides and Planning, PPG 14 (Annex 1)*. Department of the Environment and the Welsh Office, London.

Hall, J. (1999). *Uncertainty management for coastal defence systems*. PhD thesis, Bristol University, Bristol.

Hazard Forum (1996). *Hazards forum: Safety by design*. Institution of Civil Engineers, London.

Heinrichs, P. and Fell, R. (1994). Acceptable risks for major infrastructure. In Henrichs, P. and Fell, R. (eds) *Proceedings of the seminar on acceptable risks for extreme events in planning and design of major infrastructure*, Sydney, Australia. Balkema, Rotterdam.

Institution of Civil Engineers (ICE) (1999). *Safety criteria for buildings and bridges, Thomas Telford Conferences*, 1 July 1999. ICE, London.

KW Consultants (1999). *A practical model for system toughness — interim report no. 1*. KW Consultants, Health and Safety Executive, London.

Ministry of Agriculture, Fisheries and Food (MAFF) (1993). *Project appraisal guidance notes*. MAFF, London.

Ministry of Agriculture, Fisheries and Food (MAFF) (1999). *Flood and coastal defence project appraisal guidance — overview* (FCDPAG1). MAFF, London.

Ministry of Agriculture, Fisheries and Food (MAFF) (1999). *Flood and coastal defence project appraisal guidance — strategic planning and appraisal* (FCDPAG2). MAFF, London.

Ministry of Agriculture, Fisheries and Food (MAFF) (1999). *Flood and coastal defence project appraisal guidance — post project evaluation* (FCDPAG6). MAFF, London.

Nordic Committee on Building Regulations (NKB) (1978). *Recommendation for loading and safety regulations for structural design.* NKB, Oslo. Report No.36.

North Sea Coastal Management Group (NSCMG) (1999). *Flooding risk in coastal areas — an inventory of risks, safety levels and probabilistic techniques in 5 countries bordering the North Sea.* NSCMG, September (draft).

Piechowiak, A. I. (1997) *Common risk-based hydraulic design standards — a framework study.* HR Wallingford, Wallingford. HR Report SR 504.

Schoer, B. (1999). *Briefing on PPG14:1990 with Annex 1:1996 — development on unstable land.* Institution of Civil Engineers, London.

Technical Advisory Committee on Water Defences (TAW) (1997). *Safety of flood defences — a new perspective from the TAW Marsroute research programme.* Technical Advisory Committee on Water Defences, Department for Transport, Public Works and Water Management, Delft, December.

Van Gelder, P. H. A. J. M. (1999). *Statistical methods for the risk-based design of civil structures.* PhD thesis, Delft University of Technology.

Vrijling, J. K. (1990). Probabilistic design of flood defences. In Pilarczyk, K. W. (ed.) *Coastal protection.* Balkema, Rotterdam.

Appendices

Appendix 1.
Descriptions of main risk assessment techniques

A1.1. RISK METHODS AND TOOLS

A number of publications have been reviewed when compiling this description of methods. In particular, information from DEFRA (2000), the Environment Agency (2000) and Van Gelder (1999) is drawn upon extensively.

A1.1.1. Bayesian methods

Bayes theorem provides a means of using new information to revise probabilities based on old information, or, in statistical terms, to compare posterior probabilities with the priors. Where there is initial uncertainty regarding a variable, this type of analysis can be used to incorporate new information and provide new estimates, with reduced uncertainty. The methodology developed in the Institute of Hydrology (CEH, 1999), for estimating the rarity of flood peaks, provides an example of empirical Bayes estimation.

A1.1.2. Decision analysis

A process that involves the integration of utility theory, probability and mathematical optimisation to help identify the most appropriate or 'best' decision option. Initially the problem is identified and a (possibly) comprehensive list of decision options is identified. Structural analysis would organise the options into a decision tree, carefully distinguishing decision nodes (splitting points at which the outcome is chosen by a decision maker) and event nodes. Event nodes are points at which the outcome results from stochastic external events, for example the probability that a particular climate event (storm or flood level) may take place, or the probability that another decision-maker makes a particular decision, which influences one's own course of action. Next, uncertainty analysis is used to assign subjective probabilities to chance nodes, while utility analysis would stipulate cardinal utilities for outcomes.

AI.1.3. Decision analysis frameworks

Explicitly allow for (the compounding effects of) the uncertainties at each stage of the appraisal. It can be applied with an economic benefit analysis as well as non-monetary multi-criteria methods.

AI.1.4. Event trees

Event trees are used to analyse a range of likely consequences (i.e. flooding/no flooding) that may arise from a given initiating event (i.e. heavy rainfall). Fault trees work backwards from the consequence (i.e. flooding) to determine a range of possible initiating events (i.e. failure of a pump, heavy rainfall, etc.).

Both event trees and fault trees are subject to simple 'logic' rules.

AI.1.5. Expert elicitation

A range of techniques which aim to elicit information and evidence from experts on aspects of models or impacts that are otherwise difficult or not feasible to model explicitly. Expert elicitation techniques may be used, for example, to gather information about model input parameters, model processes and climate change impacts. The techniques often include methods for eliciting opinions, and also uncertainties using appropriate coding techniques, including structured questions, and graphical techniques. Techniques range from assessment by individuals (which is simple and cheap but may be less dependable) to complex structured techniques for eliciting and moderating views of groups of experts.

AI.1.6. Expert judgement

Use of evidence from individuals or groups of experts. This may relate to the likelihood of future events or scenarios, ranges or probability distributions of physical or model parameters, or possibly judgements relating to impacts, including relative benefits or disbenefits of different impacts.

This type of approach has been used extensively in asset management procedures for the dam industry, in particular a Failure Mode Element and Critical Analysis (FMECA) technique has been adopted in the recently published CIRIA guide on risk and reservoirs (Huges *et al.*, 2000).

AI.I.7. Interval analysis

A technique for assessing the effects of uncertainties on the outcome or model prediction. The aim of interval analysis is to identify the lowest possible and highest possible value of an outcome, based on extreme values of input parameters (model parameters, physical parameters, etc.). Interval analysis involves 'searching' for the combination of input parameters that together combine to produce the highest and lowest value of the output, given a particular model. Although it is conceptually simple, great care is needed to ensure that the correct combination of input values is selected. It may not be possible to select input parameter sets without trial and error, particularly for complex functions or models. Data requirements are however among the simplest of any uncertainty analysis method — only the extreme values (maximum/minimum) of all inputs considered likely or possible are needed. Of course some values may be deterministic (i.e. single values). Since the outcome is the result of *all* input parameters at their extreme values, interval analysis can give very wide bounds to outcomes. In other words, the upper and lower bounds of the output may have a very low probability of occurring. However, if the uncertainties are properly represented, the 'true' outcome is guaranteed to be within the predicted bounds. Interval analysis makes no assumptions about the probability distributions of input parameters and requires no data on this. It also does not assume any particular degree of dependence between parameters, which is partly why the resulting bounds can be so wide.

AI.I.8. Level I reliability methods

A design method in which structural reliability safety levels are specified by a number of partial safety factors, related to some predefined characteristic values of the basic variables.

A1.1.9. Level II reliability methods

A design method incorporating safety checks only at selected point (or points) on the failure boundary.

A1.1.10. Level III reliability methods

Safety checking based on exact probabilistic analysis for whole structure systems.

A1.1.11. Markov-chain modelling

A statistical mathematical modelling approach used to represent uncertainty in linked sequences of events, where each transition may be represented by a probability representing the likelihood or uncertainty that the transition will occur.

This technique has been used to extend wave data for use in the assessment of beach plan shape changes, where the temporal sequencing of storm events is an important issue. It is proposed to include this time sequence in the JOIN-SEA software developed by HR Wallingford (1998).

A1.1.12. Monte Carlo simulation

The Monte Carlo simulation is illustrated in FCDPAG4 (Figure 1; MAFF, 2000). The approach proceeds by conducting a large number of realisations of the same system model, each one with random sampling as the basis of the input parameters. By carrying out a large number of realisations covering a wide range of the input parameters, a full distribution of the output distribution is obtained.

This method is incorporated in the JOIN-SEA software (HR Wallingford, 1998), whereby many thousands of years of water level and wave height and period data can be simulated. Analytical integration of response functions, such as wave overtopping, is thus obviated, as full distributions of these response variables can be derived from the simulated data.

A1.1.13. Multi-criteria analysis (MCA)

Multi-criteria analysis is a technique for the quantitative analysis (through scoring, ranking and weighting) of a wide range of impact categories and criteria. It can encompass non-monetisable impacts and additional criteria that can be difficult to incorporate within an economic benefit cost analysis. This approach has been used by English Nature to determine the validity of transport and flood defence decisions.

AI.I.I4. Pedigree analysis

Pedigree analysis is a qualitative technique used in decision analysis to help define the state-of-the-art, expertise, credibility, potential reliability, or degree of consensus associated with information or knowledge. Hence statistical models (in general) have a lower pedigree than process-based models underpinned by theory enjoying a wide scientific consensus. Consider the consequences of sea-level rise exceeding a particular value. The opinion of a world expert in coupled climate-ocean modelling as to the likelihood of sea level exceeding that value at any particular time in the future might be given a pedigree score of 5 (out of 5). However, his or her opinion as to the consequences for a local conservation site in the UK may be of less value, and given a score of 3 (he or she is still an expert on coastal issues). In comparison, a local conservation officer with good knowledge of the potential effect of that amount of sea-level rise may have the impact opinion rated 5. The acceptability of such an impact, as determined by a survey of local stakeholders, may be given a higher pedigree than that of either the climate expert or the conservation officer.

AI.I.I5. Probabilistic risk assessment

Probabilistic risk assessment is characterised by:

- a systematic approach to failure mechanisms
- expression of loads and strengths in probabilistic terms rather than deterministic terms
- derivation of a probability of failure and comparison with some acceptable probability.

AI.I.I6. Robustness analysis

Robustness analysis may be used to help determine the robustness of the answers within an options *appraisal* to possible uncertainties as to the values of key sensitive variables and parameters (as identified from the sensitivity analysis). It identifies the extent to which the decision maker might be exposed to potential costs and errors if some uncertain eventualities regarding these parameters should arise in future.

A1.1.17. Risk register

A record of risks, their consequnces and significance and proposed management measures.

A1.1.18. Risk screening

Techniques used to identify hazards, processes and impacts which are, and are not, significant in the overall decision-making process. These are 'broad brush' techniques which generally require a reasonable understanding of the system. Screening tests are by their nature approximate and so should be designed to be conservative so that important issues are not rejected at an early stage.

A1.1.19. Sensitivity analysis

A generic term used in both formal (e.g. mathematical modelling) and in decision analysis for techniques that identify key assumptions, variables or parameters for which uncertainty as to their values could significantly affect outcomes and decisions. The technique involves examining the consequences (as determined on outputs or outcomes) of changes in the values determined for each component. For example, a decision may be sensitive to the value of discount rate used within a cost-benefit analysis. Monte Carlo techniques can be frequently used in formal analyses of model sensitivities. At each stage in an appraisal, the assessor should focus attention on those parts of the analysis or variables highlighted by analysis of sensitivity, and seek alternative and better options which could better accommodate uncertainties regarding these variables (see robustness analysis — Section A1.1.16).

A1.1.20. Statistical modelling

Statistical approaches based on an analysis of co-variation between observed climate and non-climate variables are often used as a basis for extrapolation to possible future climates. Such models (e.g. multivariate statistical models), where the causal relationships between climate and outcome variables are often highly uncertain, may have poor predictive power. Predictions from such models should always be accompanied by estimates of the confidence intervals attached to the output variables. These provide a description of the uncertainty and variability represented by the model (which itself should be regarded as uncertain). In practice, deterministic predictions are often presented, but should be avoided.

AI.I.21. Tiered risk assessment

Many issues are too complex to be calculated completely and a tiered or staged approach is more appropriate. Initial stages of the assessment aim to identify a wide range of hazards and issues that may affect a decision. These are filtered or screened to identify those which have the most important impacts. It may then be appropriate to carry out a stage of prioritisation, often using a scoring scheme to identify the most important risks. Detailed quantitative analysis can then be focussed on the key hazards and risks which are likely to be most influential on the decision. There is inevitably a degree of iteration in this approach.

AI.I.22. References

Centre of Ecology and Hydrology (1999). *Flood estimation handbook*. CEH, Wallingford. ISBN 0948540 94 X.

Environment Agency (2000). *Climate adaptation risk and uncertainty: draft decision framework*. Environment Agency, Bristol. Report No.21.

HR Wallingford (1998). *The joint probability of waves and water levels JOIN-SEA: a rigorous but practical new approach*. HR Wallingford, Wallingford. HR Report SR 537.

Hughes, A., Hewlett, H., Samuels, P. G., Morris, M., Sayers, P., Moffat, I., Harding, A. and Tedd, P. (2000). *Risk management for UK reservoirs*. CIRIA, London. Report CS42.

Ministry of Agriculture, Fisheries and Food (2000). *Flood and coastal defence project appraisal guidance — approaches to risk (FCDPAG4)*. MAFF, London.

Van Gelder, P. (1999). *Statistical methods for the risk based design of civil structures*. PhD thesis, Delft University of Technology.

Part B

Detailed case study analysis

These case studies have been selected to demonstrate selected principles that are discussed in Part A. HR Wallingford was responsible for all of the additional analysis discussed in the acceptable risk section of each case study. The overall essence of the design work undertaken by each designer has been captured and the reader should treat it as a summary of the key points rather than as a complete design report.

Case study 1.
Clacton coastal defence scheme, UK

C1.1. CASE STUDY OBJECTIVES

The objectives of the case study are to:

- describe the design process used to procure the final design
- demonstrate the principle of parallel and series systems in selecting sections of coastline (or river bank) to undertake works on
- demonstrate the principle of considering the temporal variation in risk during the lifetime of a design, using the example of the beach management scheme.

C1.2. BACKGROUND

C1.2.1. Description of site

Clacton is located on the east coast of England just north of the Thames Estuary (Figure C1.1). The area covered by this study extends for 3·8 km from the Martello Bay breakwater at west Clacton to the Bel Air roundhead at Jaywick (Figure C1.2).

The coastal defences protect an area with five main attributes:

- residential properties — there are approximately 2000 properties at risk of flooding
- commercial interests — established fishing industry which takes place close to the shore
- tourism – as a seaside resort, tourism is an important economic activity

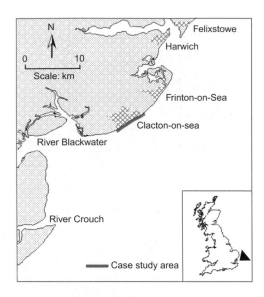

Figure C1.1. Location of Clacton sea defences

- environment — the area is a Site of Special Scientific Interest (SSSI), a Special Protection Area (SPA) and a candidate Special Area of Conversation (cSAC)
- heritage — there is evidence of Palaeolithic human occupation at the site.

C1.2.2. Development of the coastal defences at Clacton

Over the last 60 years, there has been a concentrated focus on protecting the coastline at Clacton as the site has had a number of failures (Table C1.1).

The construction of the fishtail breakwater scheme in the 1980s has been largely successful at limiting losses of sand from the frontage. However, the layout of structures did not provide a uniform level of protection along the frontage, with pinch points of low beach levels within the groyne bays. The lengths with low beach levels extend over 60% of the frontage as Figure C1.2 shows, and would cause the following problems if a 'do nothing' approach was adopted.

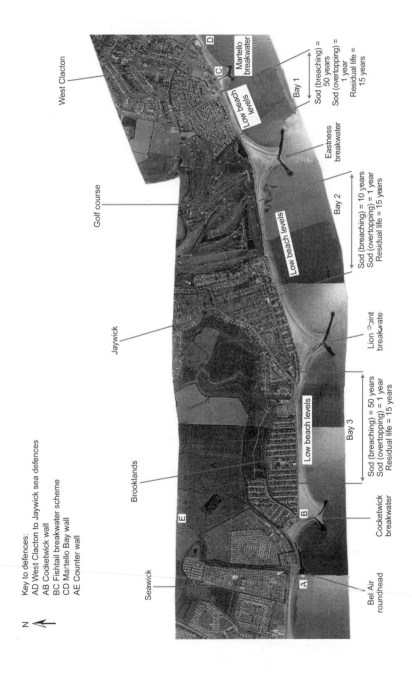

Figure C1.2. Aerial photograph of the frontage before construction of new scheme

Table C1.1.   Development of the coastal defences up to the design of the new scheme

Year	Description of events
1940s	The defence system dates from the 1940s. The original seawall comprised a reinforced concrete parapet with a sloping apron of concrete steps or revetment blocks and a steel sheet pile toe
1953	During a storm event in 1953 the seawall immediately to the west of the study frontage breached and a large area flooded resulting in the loss of 35 lives and making 600 people homeless
1972	Timber groynes, installed in 1972, had largely fallen into disrepair by 1984
1978	In 1978 an easterly storm inflicted severe damage over 100 m of the seawall at Jaywick and 320 people were evacuated
1982	Another storm event in 1982 flooded low-lying residential areas and evacuations were necessary
1980s	The falling beach level along the frontage resulted in regular overtopping of the seawall and flooding of the urban area. The seawall toe erosion was significant and this reduced the stability of the toe piling. A point was reached when urgent action was required. In this period, a review of the existing strategy of periodic improvements of the existing seawalls was undertaken. It was decided that beach renourishment was the most appropriate action rather than other more environmentally aggressive works on the seawall
1986	The Fishtail Breakwater Scheme was constructed. The scheme cost approximately £10 million and comprised beach nourishment (sand and shingle) and construction of four fishtail breakwaters and a roundhead
1990	Cocketwick Wall — the beach levels were such that it was necessary to construct a rock toe to stabilise the wall, giving a standard of defence above 150 years
1991–1998	Importation and recycling of beach. The winter of 1996/97 saw the placement of a short length of rock revetment to protect the toe of the wall at Martello Bay

- The overall residual life of the structure if a 'do nothing' approach was adopted would be approximately 15 years due to high exposure of wave attack on the seawall with reducing clay levels.

- With lower beach levels, there is a high risk of overtopping with unacceptable flooding occurring on an annual basis.
- In terms of a local structural failure due to adverse storms, the standard of protection against breaching is between 1 in 10 years and 1 in 50 years, which means that there is a significant risk of extensive flooding to West Clacton, Jaywick, Brooklands and Cocketwick.

In 1998, Posford Duviver carried out a feasibility study that claimed that the current spacing of the groynes were too wide to sustain a healthy beach (Posford, 1998). A range of solutions was examined as part of the study. These included:

- continued management of the beach with recycling and importation
- seawall strengthening and realignment
- further fishtail breakwaters
- detached breakwaters
- modifications to the existing structures.

All schemes were appraised on technical, environmental and economic merits. The solution selected and constructed was:

- extensions to the arms of the Martello Bay breakwater
- a fishtail breakwater in Bay 2
- a detached breakwater in Bay 3
- beach nourishment throughout the frontage.

The whole-life cost of the scheme over 50 years was £11 million and had a BCR of 2·56. The scheme was approved for grant aid funding by DEFRA.

C1.3. INFLUENCE OF STAKEHOLDERS ON THE ACCEPTED SOLUTION

Stakeholders at Clacton were quite diverse. Some of these stakeholders are statutory and must be consulted while others have a significant interest in the area and are still consulted but have no statutory responsibility. The main issues that affect the study area are shown in Table C1.2 with the associated stakeholders (statutory consultees are italicised).

Table C1.2.   Main issues at Clacton

Overall issue	Specifics	Stakeholders interested
Flooding	Significant overtopping or breaching of the defences	*Environment Agency* *DEFRA* Local residents Local businesses (e.g. tourist industry)
Nature conservation and ecology	Protection of designated conservation and habitat sites within and adjacent to the study area	*English Nature*
Heritage	Protection of significant finds such as the Palaeolithic site. Also protection of Martello Towers, which are designated as Scheduled Ancient Monuments	*English Heritage*
Water quality	The beaches are designated bathing beaches under the EC Bathing Directive. Also a number of outfalls that discharge near the shore	Anglian Water *Environment Agency* *Tendring District Council*
Fishing	Protection of commercial and recreational fishing	DEFRA Commercial fisherman Recreational Anglers
Ownership of the beach	Who has responsibility for the beach?	*Crown Estate* (below the high water mark) *Tendring District Council* (above the high water mark)
Tourism	Protection of amenity areas such as the sea, the beach and local golf courses	Local residents Local business
Adjacent coastlines	Protect adjacent coastlines from damage due to actions within the study area	*Tendring District Council* Shoreline Management Partnership

The stakeholders were consulted by letter with a scoping document and location map that indicated the objectives of the study and the various engineering measures under consideration, and invited consultees to contribute information to the study with their comments. Following the first round of consultation, a second letter was sent to those consultees who had shown an interest in the study, with an explanation of the preferred options under consideration.

Through this consultation process, the designer was able to build up a picture of what was acceptable and start to quantify risk levels where appropriate. Table C1.3 summarises the boundaries of acceptable risk set out by the different stakeholders.

The consultation process and the stakeholders had an influence on the outcome of the design. Local consensus stated that the original scheme constructed in the 1980s was visually acceptable and that similar structures should be used to reduce the groyne spacing to an acceptable level. The preferred design proposed that fishtailed groynes should be constructed in the midpoint of each bay. Discussions with English Nature established the value of the ancient river bed system with Palaeolithic remains that would underlie the proposed fishtail breakwater in one of the bays. This established value deemed the original design to be unacceptable and the alternative detached breakwater was proposed and accepted. If this issue had not been established in the early stages of the detailed design then the consequences could have been more significant with substantial changes having to be made during construction or resulting in the loss of an important archaeological site.

C1.4. DESIGN PROCESS

C1.4.1. Identification of need

The Environment Agency as the client had indicated that action was required at Clacton with the ongoing process of beach management and increasing perception that the risk of flooding was increasing. However, the identification of a need was part of a bigger process that had started with a regional overview of the local coastline through a shoreline management plan where prioritisation was given to lengths of coastline. Through this process, work had been completed first to the west of the study area between Jaywick and Colne Point. The selection of one site as opposed to another is discussed later on in the case study.

Table C1.3. Acceptable risk matrix for Clacton sea defences

Stakeholder	Unacceptable	Tolerable	Broadly acceptable
English Nature	Any impact on designated areas, especially a local bird breeding area	Do nothing	Improvement to the beach to maintain local bird breeding area
	Hard engineering solution over any archaeological SSSI	Do nothing on the beach	Soft engineering solution that has minimal impact on archaeological find
English Heritage	Any damage to the Martello Towers	Reduce risk to Martello Towers	No damage to Martello Towers
Tendring District Council (TDC) and the local community	Reduction of beach quality and amenity value area	No further deterioration of the beach	Improvement of beach
	A high risk of extensive flooding	No reduction in standard of flood defence	A high standard of flood defence
Crown Estate	Damage to the beach	No damage	Improved management of the beach
Anglian water, Environment Agency, TDC	Quality bathing water reduced	Do nothing	Quality bathing water improved
Anglian Water	Outfalls damaged by any works	Do nothing	No damage to outfalls
Environment Agency	Standard of defence < 1:200 years	Standard of defence 1:200 years	Stand of defence > 1:200 years
DEFRA	Restriction to fishing		
	BCR less than 1 over 50 years	BCR between 1 and 2 over 50 years	BCR greater than 2 over 50 years

During this stage, the designer set about assessing historical data, the current condition of the flood defences and collecting data that would be used in the design, such as wave and water level data.

C1.4.2. Functional analysis

The designer then moved on to setting out the scheme objectives through discussions with stakeholders, as shown in Table C1.3. The other stage was to set out the design criteria against which each design was to be tested. The main design criteria used were as follows.

- Design life — British standards state that the design of maritime structures should have a design life of 60 years. This is only guidance and in this particular case the design life of the scheme was taken to equal the economic life of the scheme as stated by DEFRA, which was 50 years.
- Standard of protection — DEFRA publish in their *Flood and coastal defence project appraisal guidance notes* (MAFF, 1993) a set of indicative standards for flood and coast protection. The designer assessed that the hinterland at Clacton was a medium density urban area, which suggests that the indicative standard of protection should have a return period of 150 years.

  However, this standard is only indicative and the true test of protection required is based on economics and the BCR. In this particular case, the Environment Agency were seeking Government funding and therefore needed to satisfy DEFRA's grant aid procedures, which state that for a scheme to be acceptable the BCR must exceed 2 over a period of 50 years. DEFRA dictate that during the economic assessment the discount factor used is 6% and assume that the residual value of the scheme at the end of 50 years is zero.

  The standard of protection is tested against sea level rise, which is also dictated by DEFRA as being 6 mm/year.
- Other scheme objectives such as environmental, heritage and amenity was also included and considered through contact with the stakeholders.

C1.4.3. Generate alternative solutions

A number of options were considered by the consultant including:

- continued management of the beach with recycling and importation
- seawall strengthening and realignment
- further fishtail groynes
- detached breakwaters
- modifications to the existing fishtailed groynes.

C1.4.4. Comparison and selection

The designer assessed each option against the design criteria and three schemes were proposed for further analysis.

1. Detached breakwaters with beach renourishment.
2. Detached breakwaters with beach renourishment and modifications to the existing fishtail breakwaters.
3. New fishtail breakwaters with beach renourishment.

To make further comparison and selection, all three schemes were tested using a combination of numerical and physical modelling (HR Wallingford, 1998).

From the information obtained through the modelling, robust costs were developed and the assessment of the likely economic benefits of the scheme made (i.e. the 'do nothing' damages minus any residual losses). The most economically viable option was the third option, new fishtailed groynes between the existing groynes. This scheme was accepted by the stakeholders at the time and moved onto the detailed design and specification stage.

C1.4.5. Detailed design and specification

Once the plan shape of the scheme had been selected during the comparison and selection process, the actual detailed design and specification was relatively simple as the designer had the benefit of being able to use the previous engineering structures as the basis. The consequence of this was that uncertainty was perhaps not as high compared to working from a blank piece of paper. Checks were made on the stability of the rock groynes using up-to-date guidance and the renourished beach material had the same specification as previously.

A major impact on the design process was observed through English Nature who would not permit the construction of a fishtailed groyne as it came over an archaeological site. The design had to be changed and a detached breakwater was used instead in Bay 3. The consequence of this late design change was minimal because work on detached breakwaters had been undertaken previously during earlier stages. However, further late information could have had a significant negative impact on the project.

C1.4.6. Construction

No major changes to the design were made during the construction phases.

C1.4.7. Management

It is estimated that renourishment to counter losses of beach material (assessed at 10% over ten years from the modelling) would be undertaken at ten yearly intervals. Some minimal maintenance on the groynes and the seawall are expected over the economic lifetime of the scheme.

C1.4.8. Decommission

The designer expects that the scheme will still be functional after 50 years and therefore will not be likely to be decommissioned.

C1.5. ACCEPTABLE RISK ISSUES

This section of the case study was undertaken solely by HR Wallingford as part of the research project. Analysis has been used to demonstrate the basic principles set out in this book and should not be considered as accurate design data for other projects.

C1.5.1. Application of the Acceptable Risk Bubble

Conceptually, the Acceptable Risk Bubble is a powerful tool to help the designer manage multiple risks efficiently. In practice, this thought process might be used by a number of designers already but there is very little information available on the key steps required to actually manage a multi-attribute problem. This case study illustrates how such a process could work in practice and demonstrates the benefit of the analysis.

The main steps that are considered are:

- identify stakeholders and risk owners
- brainstorm risks/concerns and rank them
- select top level concerns
- define value curves for each concern
- define broadly acceptable, mid-tolerable and unacceptable levels
- score projects against Acceptable Risk Bubble.

The possible process is now discussed below.

Identify stakeholders
Stakeholders identified were:

- Posford Duvivier (designer)
- Environment Agency (client)
- DEFRA
- local community (residential and commercial)
- English Heritage
- English Nature
- Anglian Water
- Tendring District Council
- other local maritime authorities.

The key risk owners were:

- Posford Duvivier (designer)
- Environment Agency (client)
- DEFRA
- English Heritage
- English Nature
- Anglian Water
- Tendring District Council.

Brainstorm risks/concerns, rank them and select top level risks/concerns
The following risks/concerns were highlighted as being key to the success of the project:

- engineering
- economics
- flood defence

- nature conservation and ecology
- heritage
- water quality
- fishing
- tourism
- impact on adjacent coastlines.

Define broadly acceptable, mid-tolerable and unacceptable levels
The value curves for each risk/concern have been tabulated qualitatively, see Table C1.4.

Score projects against Acceptable Risk Bubble
The criteria for scoring projects were based on each of the risk categories as:

- broadly acceptable equal to or less than 1
- tolerable between 1 and 3 (mid-point of tolerability equals 2)
- unacceptable equal to or greater than 3.

The scoring estimates for the project are presented in Table C1.5 and the results plotted onto the Acceptable Risk Bubble (Figure C1.3). The analysis shows that the project was reasonably balanced but some work had to be done to make sure the heritage and environmental stakeholders were satisfied with the final solution.

C1.5.2. Distribution of risk levels

Designing a structural system so that there are no obvious weak links is a clear goal of any designer, if there is a belief in the analogy that the chain is only as strong as its weakest link. In the vast amount of cases in coastal and fluvial engineering the designer is refurbishing an existing structure. Where the existing structure is ascertained to be a weak link, it may be better to keep that existing structure but instead also provide more protection to the existing structure (e.g. raise the beach levels in front of an existing seawall). There is, however, a clear need to gain a broader understanding of what designers are currently doing and where they can focus efforts to improve the efficiency of their designs. The case studies have been used to illustrate the risk levels accepted for each structural element and how these risk levels potentially interact as part of the overall system.

Table C1.4. Clacton — acceptable risk matrix

Risk/concern	Broadly acceptable	Mid-tolerable	Unacceptable
Engineering	Installation of groynes and refurbishment of the seawall	Additional groynes to maintain a constant width beach right along the frontage	Removal of existing groynes and a complete change in solution
Economics	High BCR	BCR equal to 1	Gain no economic benefit from undertaking works
Flood defence	Wave-induced overtopping limited to spray causing no flooding in the area	Limited wave-induced overtopping during storms with a annual return period of 200 years	Significant wave-induced overtopping or breaching during an event with an annual return period less than 200 years
Nature conservation and ecology	No damage to designated conservation and habitat sites within and adjacent to the study area	No damage to designated conservation and habitat sites within and adjacent to the study area	Any damage to designated conservation and habitat sites within and adjacent to the study area
Heritage	Full protection of significant finds in the area	Full protection of significant finds in the area	Damage to significant finds in the area
Water quality	No negative impact to the quality of bathing waters	No negative impact to the quality of bathing waters	Negative impact to the quality of bathing waters
Fishing	Full access for commercial and recreational fishing	Status quo	No access for commercial and recreational fishing
Tourism	Enhancement to the tourism industry	Status quo	Negative impact on tourism, loss of beach and coast path
Impact on adjacent coastlines	Positive impact to the management of adjacent coastlines	No impact	Negative impact to the management of adjacent coastlines

Table C1.5.   Clacton — acceptable risk scoring table

Risk/concern	Score	Reason
Engineering	2·0	Tolerable position
Economics	1·8	Economically viable, but not a very high BCR
Flood defence	2·0	Tolerable position
Nature conservation and ecology	2·5	Design attempting to control habitat, so some residual damage is likely
Heritage	2·5	Change in design to accommodate not building a structure directly on top of an ancient river bed, but structures still being constructed so some residual damage is likely
Water quality	2·5	Outfalls still discharging, residual risk of beaches being contaminated
Fishing	2·5	No change to current situation
Tourism	1·8	Improved beach
Impact on adjacent coastlines	1·9	Limited positive impact on adjacent sections of coastline

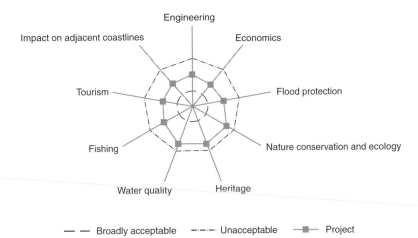

Figure C1.3.   Clacton — Acceptable Risk Bubble

The main steps in the case study analysis were:

- identify key structural elements of the system
- identify the design discipline they are affiliated to
- identify their reliance on other structural elements
- define the lifetime probability of failure accepted for each element.

The coastal defence at Clacton can be subdivided into three main structural elements, as follows.

- *Beach control structure (groyne or breakwater)*. The beach control structure is designed to resist hydraulic conditions up to an annual return period of 150 years with minimal maintenance over the structure's 50-year working design life. The structure sits mainly within the hydraulic discipline, although there is some interaction with geotechnics to ensure there is no unacceptable settlement of the structure over time. The beach control structure does not rely on any other structural element to maintain its own integrity.
- *Beach*. The beach is designed to resist hydraulic conditions up to an annual return period of 150 years. The movement of beach material is controlled by the beach control structure that limits material leaving each groyne bay. However, maintenance every ten years is incorporated into the design in order to maintain a beach width that will resist the design conditions. The reliability of the beach system is a function of the beach itself and the beach control structures. The beach is designed within the hydraulic discipline.
- *Seawall*. It is assumed that the 15-year residual life of the existing seawall is prolonged by protection from the beach. The reliability of the seawall is a function of the seawall itself, the beach and the beach control structures.

The crude analysis of accepted risk levels in terms of quantifying the lifetime probability of failure over the design working life shows a wide range of risk levels accepted for each individual element (Table C1.6). When the elements are considered as part of the overall system, a large reduction in the individual risk level of the seawall over the 50-year working design life of the system is observed due to the protection provided by the beach (Table C1.7). In this particular case study, the primary design discipline controlling the design is hydraulics with some geotechnical interaction as part of the beach control structure design.

Table C1.6.   Clacton — accepted risk levels for each structural element

Structural element	Discipline*	Design return period: years	Working design life: years	Individual lifetime probability of failure: %†
Beach control structure	H/G	150	50	28
Beach	H	150	10	6
Seawall	S	15	50	96

* H = Hydraulics; G = Geotechnics; and S = Structures.

†$P_{individual} = 1 - (1 - 1/TR)^n$ where TR is the design return period and n is the design life.

Table C1.7.   Clacton — accepted risk levels for each structural element as part of a system

Structural element	Reliability function	Total lifetime probability of failure: %*	Reduction in lifetime probability from individual to total: %†
Beach control structure	Beach control (P_1) Structure (P_2)	28	0
Beach	Beach control (P_1) Structure (P_2) Beach (P_3)	2	4
Seawall	Beach control (P_1) Structure (P_2) Beach (P_3) Seawall (P_4)	2	94

*$P_{individual} = 1 - (1 - 1/TR)^n$ where TR is the design return period and n is the design life.

†$P_{total} = P_1 \times P_2 \times \ldots$ and so on.

C1.5.3. Selection of one piece of coastline to focus on against another

In the area of coastal and flood defence, funders have to make informed decisions about where to do work first and where to wait. In the UK, DEFRA use a priority score that sets out which schemes require funding first. This priority assessment is based primarily on economic assessments, but also by environmental or social policy objectives. Where government funding is not involved, the decision to fund works may be due to a wide range of reasons, such as a private landowner wishing to make improvements to his or her property.

Clacton is an interesting case study because the natural flood zone is split into two parts, the study area and the 'other area' protected by the adjacent coastline west of the study area between Jaywick and Colne Point. A secondary flood defence separates the study area from the adjacent 'other' area. The secondary defence is an earth embankment that extends perpendicular to the coastline. If one of the two frontages fails then the secondary defence protects the area on the opposite side of the secondary defence. During the selection of which frontage to focus on first, this secondary defence was assessed to ensure that if flooding did occur on one side then it would be contained within its own flood area. This system can be described as running in parallel because failure of one component does not impact on the other (Figure C1.4).

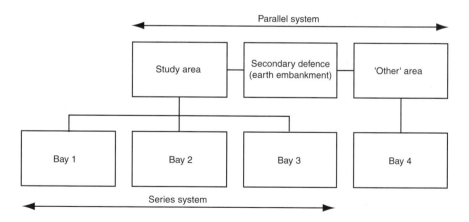

Figure C1.4.   Series and parallel systems

However, the Clacton frontage is split into three main bays (i.e. the seawall between each fishtail groyne). If the seawall in one of bays fails then the whole system will fail because they are connected to the same flood zone and therefore failure has the same consequence. This system can be described as running in series because it only takes one of the defences to fail to cause an impact on the whole flood zone (Figure C1.4).

When designing a system that runs in series, there is little value in designing all of the defence lengths to different risk levels as it is the weakest link in the series that will cause failure. Funders, such as DEFRA, will wish to fund a scheme that encompasses all of the elements that may lead to failure. Care needs to be taken by the designer to ensure that the system, its links and what are the consequences of a partial/total failure are properly understood. This will ensure that a design issue is not overlooked and any weak links in a design are identified.

C1.5.4. Temporal variation in risk levels

The designer considered that there were five potential failure mechanisms that could cause catastrophic failure of the defences:

- excessive overtopping causing structural damage to the rear face of the seawall
- instability at the toe of the seawall, caused by undercutting of the clay foundation
- breach of seawall due to long-term exposure to wave attack, overtopping and fatigue of the structure
- piping underneath the seawall
- overturning of the seawall.

The risk of the last two occurring are intuitively low compared to the first two because:

- the concrete seawall is sufficiently robust in the short term to withstand direct wave attack
- the wall is founded on clay reducing the seepage of water under the structure
- the concrete seawall is sufficiently bulky that overturning is very unlikely.

The designer focused therefore on the possibility of excessive overtopping and instability at the toe of the seawall. To reduce the risk of these failure mechanisms occurring, the designer made the decision to renourish the beach so that overtopping was controlled and the toe of the seawall was not exposed. The design of this beach consisted of three main elements, each with different risks (Table C1.8).

The design risks were managed by providing a beach that protected the seawall and stopped seawater entering the flood zone. The beach was designed to resist a loading condition (i.e. sea conditions) which was representative of that which occurs at least once in 150 years. The same level of safety was given to the seawall without any major works because the beach protected the seawall from further deterioration; the structural integrity of the seawall relies on the beach being in place. If the beach fails then the risk level of the

Table C1.8.   Design standards used for each element and associated risks

Design element	Design return period	Risks to performance
Groyne	1 in 150 years	Collapse of groyne. Risk of collapse unlikely to vary significantly over time because rock armour durable and minimum maintenance required
Beach	1 in 150 years	Loss of beach material that increases wave loading to the seawall sufficient to induce an increase in overtopping and ultimately a catastrophic failure of the seawall. The deterioration of the beach is controlled by the performance of the groynes
Seawall	1 in 150 years	Loss of beach material so that beach levels reduce to a level where overtopping and wave attack increase the wave loading. High exposure would mean it is likely that the seawall would collapse within 10–15 years, so the residual strength of the seawall has an approximate residual life of 1 in 10 years

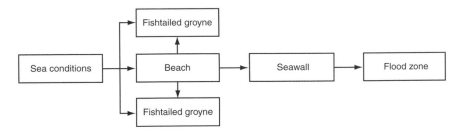

Figure C1.5. Reliance diagram on water entering the flood zone

flood protection system will increase to the current residual strength of the seawall. The risk of beach movement was controlled by the construction of new groynes between the existing ones with the aim of curtailing any excessive fluctuations in beach width and loss of beach material. The groynes were designed to structurally resist a 1 in 150 year coastal storm without the beach in place to protect them and are therefore independent of the beach itself, but the beach relies on the groynes to reduce beach material being lost. The designer predicted a 10% loss of beach material over ten years and, therefore, the beach would require renourishment near the end of this period. The reliance of the different elements is illustrated in Figure C1.5 (reliance is dictated by the direction of the arrow, e.g. the beach is reliant on the beach control structure).

The performance of the beach is, therefore, critical to the success of the scheme but it is also holds the most uncertainty in terms of its performance. Once the beach is placed and then left, the standard of protection afforded by the beach will reduce as beach material is lost and the beach width retreats towards the seawall at pinch points along the frontage.

In terms of design, the standard of protection was set at 150 years but practically the risk level fluctuates as the beach width increases or decreases through time along the frontage (i.e. one section will increase in beach width as the other decreases). The designer made the judgement that the increase in risk along some parts was not significant if monitored and managed appropriately. However, if the beach continued to reduce then some mitigation measures such as renourishment would be required. Although the 150-year profile is the target standard, there is an acceptance by the client that this may fluctuate and there may be some lengths of the beach where widths are reduced and the target profile does not comply. The solution was to monitor and maintain when required, which demonstrates the importance of an effective and robust beach management plan to address these risks actively.

This case study demonstrates that each element in the design must be investigated thoroughly so that the risks are clearly identified not only in their current state but also as these change over time. Different elements will deteriorate in different manners and the method of managing the risk will need to be appropriate for the given situation. Here the designer decided to reduce fluctuations in the beach width by decreasing the groyne spacing and setting out a clear beach management plan based on modelling results rather than to constantly renourish every year or to build a new seawall.

Another approach to the beach design may have been to design to a higher standard (e.g. 1 in 1000 years) and then allow the system to degrade over time so that the standard of protection is 1 in 150 years at the end of the design life. In this case, it is likely that such an approach would have been excessively expensive and therefore economically unsustainable. However, in other contexts such an approach might be appropriate. For example, the design of steel sheet piling, say, increasing the thickness of steel to resist abrasion so that the pile lasts for 50 years, may be more cost-effective than replacing a steel sheet pile every ten years. A temporal understanding of the risk is, therefore, a very important risk parameter in terms of not only structural safety but also for the whole-life cost of the scheme.

C1.6. CONCLUSIONS

The main conclusions of the case study are as follows.

- The accepted design was influenced by the political pressure to have similar structures to those constructed in the 1980s. The heritage of the site also dictated a change in the design from a fishtailed groyne to a detached breakwater in one bay. This heritage requirement was not indicated during the early stages of the design process and a late design change could have had a major financial or time consequence on the project.

- The detailed design was focused on the performance of the beaches through detailed modelling while the design of the groynes and seawall were based on previous experience of the existing structures on the site. Design checks were made to ensure that structures complied with current best practice.
- The decision to do works at a certain section of coastline or riverbank needs to consider the risks and the interaction of individual elements that could impact on that risk. Care needs to be taken that no element is ignored that may cause the same risk to still occur.
- The nature of risks can change over time and the temporal changes need to be properly understood so that sufficient durability can be built into the design by either building more strength into the structure or installing a strict inspection and maintenance regime.

CI.7. ACKNOWLEDGEMENTS

HR Wallingford would like to thank Posford Duvivier and the Environment Agency for providing the time and data required in this case study. Special thanks go to Dr Noel Beech, the principal contact at Posford Duvivier.

CI.8. REFERENCES

HR Wallingford (1998). *Clacton model studies*. HR Wallingford, Wallingford.

Ministry of Agriculture, Fisheries and Food (1993). *Flood and coastal defence project appraisal guidance notes*. MAFF, London.

Posford Duvivier (1998). *West Clacton to Jaywick sea defences*. Detailed Appraisal Report (unpublished).

Case study 2.
Castle Cove coast protection

C2.1. OBJECTIVES OF THE CASE STUDY

The objectives of the case study are to:

- describe the design process to develop the final design
- demonstrate differences and benefits between deterministic and probabilistic design methods.

C2.2. BACKGROUND

C2.2.1. Description of the site

Castle Cove is a 250 m section of coastline on the south coast of the Isle of Wight, just off the south coast of England (Figure C2.1). The area of interest is located to the west of Ventnor, a town with 6500 inhabitants.

The majority of the coastline along the south coast of the Isle of Wight lies within an ancient landslide complex, known as the Undercliff, in which the town of Ventnor and Castle Cove are located. The Undercliff extends for a distance of approximately 12 km along the coast and 500 m inland. The rear scarp of the landslide reaches an elevation of +170 m OD and offshore investigations have confirmed that the landslide extends in excess of 1·5 km beyond the current shoreline. Coastal erosion has been identified as the principal factor in the long-term instability of the Undercliff. Erosion undermines the toe support of the landslide and promotes ongoing movement.

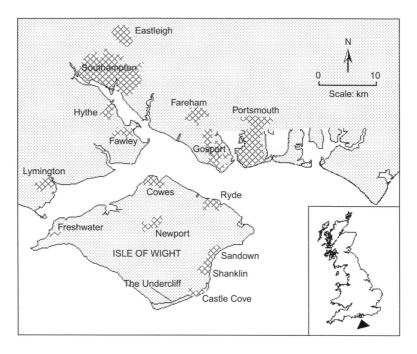

Figure C2.1. Location of Castle Cove

At Castle Cove, the ongoing marine erosion of the coastal slopes and their consequent recession was threatening the stability of the properties situated on and immediately adjacent to the coastal slopes. Major slope failures occurred that destroyed footpaths situated on the coastal slopes and significantly increased the instability of surrounding properties and the area in general. The slopes were also recognised as being a habitat for rare and important flora and fauna.

C2.3. THE DEVELOPMENT OF COASTAL PROTECTION AT CASTLE COVE

The original coast protection measures, which were constructed in the early 1970s, consisted of a wave wall that comprises rockfill and concrete cubes within a timber and steel grillage. The wall, which was located approximately 10 m seaward of the toe of the slope prior to construction, dissipated wave energy while allowing the sea to pass through it, eroding the toe of the slope.

Continuous wave action damaged the timber grillage and much of the core material of the wall was lost or deposited on the beach. Some of this material was subsequently placed along the toe of the slope in an attempt to protect the toe from further erosion. A slope failure in January 1994 made the mudslide and landslipped material collecting at the toe of the slope susceptible to marine erosion, resulting in further over-steepening of the clay slopes and further destabilisation of the coastal slope.

The form of the existing coast protection structure was considered unfavourable with respect to the retention of fine-grained beach deposits. Prior to the construction of the original works, a natural sandy beach had existed in the area. This was largely absent except under favourable sea conditions. This depletion of the superficial beach deposits, which exposed the underlying geology to marine erosion, was considered to be due to the reflected wave energy from the structure scouring away the beach deposits.

The consequences of continued marine erosion in the Castle Cove area was considered to be twofold. On a local scale, as experienced during January 1994, erosion and instability resulted in the further recession of the coastal slopes which put at risk the residential properties, accesses, footpaths and services that occupy the cliffs and intermediate areas. Secondly and, potentially more significantly, it resulted in the loss of material from the lower sections of the Undercliff landslide complex. The observations of ongoing movement in the area indicate that this mechanism was already at its limiting equilibrium. Therefore, the loss of any further material would adversely effect the stability of the area and may result in an increase in the rate of ground movements. Even more damaging, loss of toe material could have triggered an extension of the instability with possibly significant consequences to the town of Ventnor and put at risk a significant proportion of the estimated £7·2 million of development.

The primary objective of the scheme was, therefore, to establish effective coast protection measures to prevent erosion of material from the toe of the coastal slope. This maintained the current level of toe support within the overall landslide system and also limited the rate of local instability. In addition, the implemented scheme stabilised the slope and protected safety, safeguarded property and reinstated the coastal footpath that was lost during landsliding.

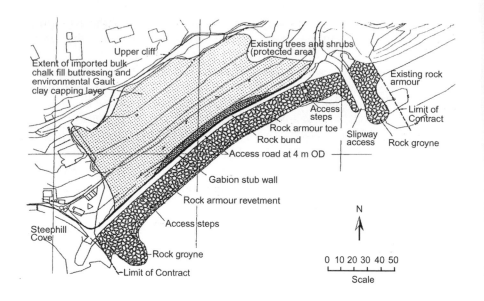

Figure C2.2.   General arrangement

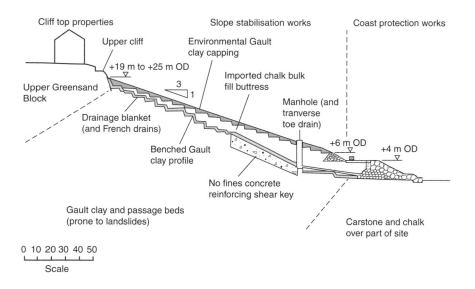

Figure C2.3.   Typical section through slope works

The scheme comprised the following main elements (see Figures C2.2 and C2.3):

- a rock armoured toe revetment to protect against marine erosion
- stabilisation of the slope through reprofiling together with slope buttressing using imported fill and a reinforcing shear key
- deep and shallow drainage measures
- an access promenade to the rear of the revetment constructed along the full length of the works
- steps and a slipway to enhance the amenity value and to enable any necessary maintenance works to be undertaken on the revetment and on the foreshore.

C2.4. INFLUENCE OF STAKEHOLDERS ON THE DESIGN PROCESS

South Wight Borough Council (SWBC), the coastal authority for the area, was the main driver of the project with the focus on protecting properties along the cliff top. However, there were a number of other interested parties, summarised in Table C2.1.

Other interested parties were Southern Water, Crown Estate Commissioners, Isle of Wight County Council, owners of adjacent properties and the Isle of Wight Joint Planning Technical Unit.

C2.5. DESIGN PROCESS

C2.5.1. Identification of a need

Landslide activity in Ventnor has been extensively reviewed and monitored over the last few years. Studies have established a relationship between the magnitude of rainfall and observed landslide activity and a conceptual model for landsliding in Ventnor (Lee *et al.*, 1998).

Extensive sections of the lower part of Ventnor are experiencing ongoing seaward movement at a rate of 2–3 mm a month along shear planes within the clay layers of the Sandrock. Major failures occurred in 1912, 1960, 1961 and most recently in 1994, following a period of continuous rainfall (Clark and Fort, 1994).

Table C2.1. Main stakeholders and their areas of interest

Stakeholder	Area of interest
South Wight Borough Council*	Local coastal authority responsible for coastal protection
Ministry of Agriculture, Fisheries and Food	Providing grant aid to the scheme. Interested in value for money and protection of the environment
English Nature	Interested in protecting the rare flora and fauna that exists on the coastal slopes. The site is recognised as being a Special Area of Conservation, proposed Site of Special Scientific Interest (SSSI) and an Area of Outstanding Natural Beauty (AONB)
Local community	Interested in boosting the local amenity, the reopening of coastal paths and improving emergency access to properties

* South Wight Borough Council is now replaced by the Isle of Wight Council.

With these observations and the potential consequence of £7·2 million of development being lost, it was felt necessary to find a solution for Castle Cove given the fact that the remainder of the coastal protection measures in adjacent areas were operating effectively.

C2.5.2. Functional analysis

Through consultation with the main stakeholders, the main objective of the scheme was to establish effective coast protection measures to prevent erosion of material from the toe of the coastal slope and thus retain landslide debris within the overall landslide system.

The following design parameters were set out.

- *Design life* — BS 6349 Maritime Code (BSI, 2000) states that maritime structures should have a design life of 60 years. This is guidance only and in this particular case the design life of the scheme was taken to equal the economic life of the scheme as stated by DEFRA, which was 50 years.

- *Standard of protection* — DEFRA publish a set of indicative standards for flood and coast protection in their *Flood and coastal defence project appraisal guidance notes* (MAFF, 1993). The designer assessed that the hinterland at Castle Cove was a low density urban area, which indicates that the indicative standard of protection should have a return period of 100 years. However, the true test of protection required is requested to be based on economics and the BCR. In this particular case, the South Wight Borough Council were seeking Government funding and therefore needed to satisfy DEFRA's grant aid procedures, which at that time the BCR must exceed 1 over a period of 50 years. The assumption was made that the residual value of the scheme at the end of 50 years is zero.

  The standard of protection was tested against sea-level rise, also set out in DEFRA procedures as being 6 mm/year for the south of England.
- *Design events* — two design events were assumed for design of the coastal protection, an ultimate limit state with a return period of 100 years and a serviceability limit state of safe public access for a return period of one year.
- *Factor of safety for slope stabilisation* — the British Standard for Earthworks (BS 6031; BSI, 1983) suggests that for slopes with a good standard of investigation, the factor of safety should lie between 1·3 and 1·4. The factor of safety selected by the designer was 1·3, based on the level of investigation and testing carried out at the site.
- *Environmental protection* — during the design process, it was recommended that a Gault clay capping layer be placed over the chalk fill buttress to encourage the reintroduction of natural local flora on the coastal slopes. The details of this proposal for environmental mitigation followed discussions with the designers and the South Wight Borough Council County ecologist and fell in line with English Nature requirements to protect the rare flora and fauna.
- *Amenity* — this was to be improved by providing a means of access to the foreshore area, which can also act as an emergency vehicular access to Steephill Cove from the west. Currently, there is no vehicular access to the properties situated in Steephill Cove.

C2.5.3. Generate alternative solutions

The consultant split the problem into two distinct elements, one being coast protection and the other being slope stabilisation. Several of the alternative schemes were put forward, based on engineering experience of the site and similar problems.

The solutions developed for coast protection included:

- do-nothing
- maintain current coastal protection
- beach recharge
- offshore breakwater
- concrete mass wall
- anchored vertical wall
- stepped concrete wall
- hollow block revetment
- concrete block revetment
- gabion wall
- rock armour revetment.

The solutions developed for slope stabilisation included:

- deep drainage of the slope
- re-profiling by cutting back the slope crest
- buttressing the whole slope
- buttressing the lower section of the slope
- a structural solution (e.g. piles, ground anchors, etc.).

C2.5.4. Comparison and selection

Comparisons of the schemes was first done qualitatively against different criteria for both coast protection and slope stabilisation.

For coast protection, the criteria headings were:

- toe protection
- structure stability
- hydraulic performance
- aesthetics
- impact on flora, fauna
- impact on adjacent coastline
- amenity and access.

For slope stabilisation, the criteria headings were:

- stability
- cost
- minimise slope works
- sustainability
- aesthetics
- impact on flora and fauna.

Each criteria was marked by the consultant and the client under the following scoring system:

- very poor
- poor
- good
- very good
- uncertain
- not applicable

or in terms of cost for the slope stabilisation:

- low
- medium
- high
- very high.

The options for both coast protection and slope stabilisation went through this qualitative process, as shown in Tables C2.2 and C2.3 respectively.

Protection of the toe of the slope was required to prevent coastal erosion and provide additional toe weighting to support the slope. The most preferred option for the coast protection at the toe was the construction of a rock revetment that provided both good toe protection and performed well hydraulically.

The preferred solution for slope stabilisation was to buttress only the lower slope with a drainage blanket. This allowed the surface of the slope to remain as Gault clay, which satisfied the requirements of English Nature.

Table C2.2. Option assessment for coast protection

Options	Toe protection	Structure stability	Hydraulic performance	Aesthetic	Impact on flora and fauna	Impact on adjacent coasts	Amenity and access
Do-nothing	V. poor	V. poor	V. poor	Poor	Good	N/A	V. poor
Maintain existing coast protection	Poor	Poor/good	Poor	Poor	Poor/good	Good	Poor
Beach recharge	Poor/good	Uncertain	Good	V. good	Poor	Good	V. good
Offshore breakwater	Poor	N/A	Good	Good	Poor/good	Poor	Poor
Concrete mass wall	Good	Poor	Poor	Poor	Poor	Good	Good
Anchored vertical wall	Good	Poor	V. poor	V. poor	Poor	Negligible	Good
Stepped concrete wall	Good	Poor	Poor	Poor	Poor	Negligible	Good
Hollow block wall	Good	Good	Good	Poor	Poor	Negligible	Good
Concrete block wall	Good	Poor	Poor	Poor	Poor	Negligible	Good
Gabion wall	Good	Good	Good	Good	Poor/good	Negligible	Good
Rock armour	Good	Good	V. good	V. good	Poor/good	Negligible	Good

Table C2.3. Option assessment for slope stabilisation

Options slope stabilisation	Stability	Cost	Minimise slope works	Sustainability	Aesthetic	Impact on flora and fauna
Deep drainage	Poor	High	Good	Good	Good	Poor/good
Crest re-profiling	Poor/good	Medium	Poor	Good	Poor	Poor/good
Buttressing whole slope	V. good	High	V. poor	Good	Poor	V. poor
Buttressing lower slope	Good	Low	Good	Good	Good	Poor/good
Structural solution (with earthworks)	Good	V. high	Poor/good	V. poor	Poor	Poor/good

C2.5.5. Detailed design

The detailed design was undertaken using British Standards. The Maritime Code BS 6349 (BSI, 2000) was used for the design of the rock revetment, and the Earthworks Code BS 6031 (BSI, 1983) for the design of the slope stabilisation. Calculations used standard formulae and no physical modelling was undertaken. Although the two elements were designed using different codes of practice, there was interaction between them to make sure that the toe protection provided adequate protection to the coastline and structural support to the slope. A diagram of the simplified process of the design is presented in Figure C2.4.

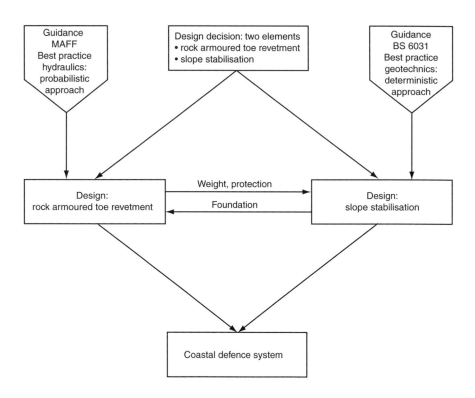

Figure C2.4. Simplified version of the detailed design process

Sensitivity of the slope stabilisation to the variability of the material strength was achieved by reducing the shear strength of the Gault clay from 25 degrees to 20 degrees, ensuring that the factor of safety exceeded unity and assuming residual strength based on testing.

C2.5.6. Construction

No major design changes were undertaken during the construction.

C2.5.7. Management

The works are expected to provide a consistent physical standard of protection over their life, necessitating routine maintenance. Annual routine maintenance costs have been estimated at £500 commencing from Year 5 of the scheme.

C2.5.8. Decommission

After 50 years the residual value of the structure is estimated to be zero but it is unlikely that the scheme will be decommissioned.

C2.6. ACCEPTABLE RISK ISSUES

This section of the case study was undertaken solely by HR Wallingford as part of the research project. Analysis has been used to demonstrate the basic principles set out in this the book and should not be considered as accurate design data for other projects.

C2.6.1. Application of the Acceptable Risk Bubble

Conceptually, the Acceptable Risk Bubble is a powerful tool to help the designer manage multiple risks efficiently. In practice, this thought process might be used by a number of designers already but there is very little information available on the key steps required to actually manage a multi-attribute problem. This case study illustrates how such a process could work in practice and demonstrates the benefit of analysis.

The main steps that are considered are:

- identify stakeholders and risk owners
- brainstorm risks/concerns and rank them
- select top level concerns
- define value curves for each concern
- define broadly acceptable, mid-tolerable and unacceptable levels
- score projects against Acceptable Risk Bubble.

The possible process is now discussed below.

Identify stakeholders

Stakeholders identified were:

- High Point Rendel (designer)
- South Wight Borough Council (client)
- DEFRA
- Local Community (residential)
- English Nature
- South Wight Borough ecologist.

The key risk owners were:

- High Point Rendel (designer)
- South Wight Borough Council (client)
- DEFRA
- English Nature.

Brainstorm risks/concerns, rank them and select top level risks/concerns

The following risks/concerns were highlighted as being key to the success of the project:

- engineering
- economics
- coast protection
- nature conservation and ecology
- access along the coastline
- tourism.

Define broadly acceptable, mid-tolerable and unacceptable levels

Table C2.4 was drawn up to define qualitatively the value curve for each risk/concern.

Table C2.4. Castle Cove — acceptable risk matrix

Risk/concern	Broadly acceptable	Mid-tolerable	Unacceptable
Engineering	Stable solution using most appropriate materials with a factor of safety greater than 1·3	Stable solution using most appropriate materials with a factor of safety near to 1·3	Unstable solution with a factor of safety lower than 1·3
Economics	High BCR	BCR equal to 1	Gain no economic benefit from undertaking works
Coast protection	No further erosion	Status quo	Large losses of land or property put into danger
Nature conservation and ecology	No damage to the Gault clay surface of the slope	Gault clay surface replaced after works	Loss of Gault clay surface with the consequence of destroying local flora and fauna habitat
Access along the coastline	Safe access along the coastline at all times	Safe access along the coastline with some restrictions during large storms	Status quo
Tourism	Enhancement to the tourism industry	Status quo	Negative impact on tourism, loss of beach and coast path

Score projects against Acceptable Risk Bubble
The criteria for scoring projects were based on each of the risk categories as:

- broadly acceptable equal to or less than 1
- tolerable between 1 and 3 (mid-point of tolerability equals 2)
- unacceptable equal to or greater than 3.

The scoring estimates for the project are given in Table C2.5 and the results plotted onto the Acceptable Risk Bubble (Figure C2.5).

Table C2.5. Castle Cove — acceptable risk scoring table

Risk/concern	Score	Reason
Engineering	2·5	Materials were selected using the nature conservation and ecology criteria and as such the best optimised engineering solution was not selected
Economics	2·0	BCR was 1
Coast protection	1·5	Stopped any further erosion at the toe of the structure
Nature conservation and ecology	2·0	Gault clay replaced after drainage works completed
Access along the coastline	1·5	Access to coastline improved but would need to be restricted during large storms
Tourism	1·5	Overall improvement to the area encouraging tourism

The analysis shows that the engineering part of the project had to be tailored to meet the environmental requirements of the project. If the Gualt clay layer had not been specified then perhaps the designer may have adopted a different solution.

C2.6.2. Distribution of risk levels

Designing a structural system so that there are no obvious weak links is a clear goal of any designer, if there is a belief in the analogy that the chain is only as strong as its weakest link. In the vast amount of cases in coastal and fluvial engineering the designer is refurbishing an existing structure. Where the existing structure is ascertained to be a weak link, it may be better to keep that existing structure and instead provide more protection to the existing structure (e.g. raise the beach levels in front of an existing seawall). There is, however, a clear need to gain a broader understanding of what designers are currently doing and where they can focus efforts to improve the efficiency of their designs. The case studies have been used to illustrate the risk levels accepted for each structural element and how these risk levels potentially interact as part of the overall system.

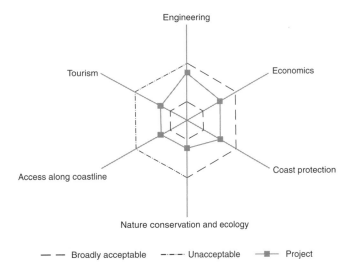

Figure C2.5. Castle Cove — Acceptable Risk Bubble

The main steps in this analysis of the case study analysis were to:

- identify key structural elements of the system
- identify the design discipline they are affiliated to
- identify their reliance on other structural elements
- define the lifetime probability of failure accepted for each element.

The coastal slope at Castle Cove can be subdivided into two main structural elements, as follows.

- *Rock revetment*. The rock revetment is designed to resist a hydraulic loading with an annual return period of 100 years with minimal maintenance over the structure's 50-year working design life. The structure sits within the hydraulic discipline. The reliability of the rock revetment is a function of the rock revetment itself and requires no other structural element to maintain its own integrity.

- *Coastal slope.* The design of the coastal slope is a complicated system to analyse and, therefore, even for engineers specialised in this field of engineering it is still very difficult to quantify a lifetime probability of failure with any certainty. The structure itself was designed to a factor of safety of 1·3 with key assumptions made on the soil and drainage parameters. The estimated lifetime probability is approximately 0·06, assuming that the toe is not in place (see Section C2.6.3 for further details), which equates to an annual design return period of 1500 years, assuming the working design life is 60 years. The design of the coastal slope sits within the geotechnical discipline. The structure itself relies on the rock revetment to provide toe support, but failure is more likely to be achieved through heavy rainfall on the slope rather than through impacting waves on the toe from the sea.

The crude analysis of accepted risk levels in terms of quantifying the lifetime probability of failure over the design working life shows a wide range of risk levels accepted for each individual element (Table C2.6). When each element is considered part of the overall system, no real reduction (only a 2% reduction for the coastal slope) in the system's risk level is observed due to the coastal slope remaining structurally sound, even with the rock revetment removed (Table C2.7). In this particular case study, the primary design

Table C2.6.   Castle Cove — accepted risk levels for each structural element

Structural element	Discipline*	Design return period: years	Working design life: years	Individual lifetime probability of failure: %†
Rock revetment	H	100	60	45
Coastal slope	G	1500	60	4

* H = Hydraulics; and G = Geotechnics.

†$P_{individual} = 1 - \left(1 - 1/TR\right)^{n}$ where TR is the design return period and n is the design life.

Table C2.7.   Castle Cove — accepted risk levels for each structural element as part of a system

Structural element	Reliability function	Conditional lifetime probability of failure: %*	Reduction in individual to conditional lifetime probability: %†
Rock revetment	Rock revetment (P_1)	45	0
Coastal slope	Rock revetment (P_1)	2	2
	Coastal slope (P_2)		

$*P_{individual} = 1 - (1 - 1/TR)^n$ where TR is the design return period and n is the design life.

$†P_{total} = P_1 \times P_2 \times \ldots$ and so on.

discipline controlling the design is geotechnics, with hydraulic interaction coming not from sea conditions but from rainfall and the saturated strength of the slope. Overall, the hydraulic element of the design had little impact on the ultimate strength of the system but was required to satisfy other design criteria such as protection of a new walkway at the base of the slope along the coast.

C2.6.3.   *Different approaches to design*

To design the coastal slope, static design assumptions were made regarding the water level within the slope, soil properties, sea conditions, etc., and a profile was developed that provided an overall factor of safety of 1·3. This design approach makes the assumption that given the magnitude of the design events selected and the factor of safety achieved, the coastal slope should remain structurally sound for its design life. Using this approach, the designer does not need to quantify what the annual probability of failure may be and predict the residual risk remaining in the scheme. However, to ensure that the probability of the slope failing was minimal, the designer included sensitivity checks on the factor of safety by considering such things as reduced soil strength or a higher water table.

As in conventional geotechnical design, to assess the probability of failure requires a rigorous approach and a clear understanding of the failure modes and interactions between design elements. The principal methods for generating probabilistic predictions include regression analysis, stochastic simulation, repeated realisations of physical process models and the structured use of subjective probabilities by experts. A probabilistic approach, therefore, relies on engineering judgement and interpretation as in the deterministic approach. The following is a simplistic attempt to predict the indicative annual probability of failure of a coastal slope and compare it with the factor of safety obtained.

The coastal slope at Castle Cove has two main failure mechanisms (Figure C2.6), which are as follows.

- Branch I — a structural failure of the rock revetment providing toe support. The coastal slope would then have a similar level of stability as if the project had not been constructed. The probability of Branch I occurring is conditional on the toe being removed, sufficient rain to trigger a failure and then the actual slope failing.
- Branch II — failure caused by high groundwater levels that create sufficient instability in the slope to cause a failure, even with the additional toe support from the rock revetment. The probability of Branch II occurring is conditional on the toe remaining, excessive rain to trigger a failure and then the actual slope failing.

Both of these branches are discrete events with the probability of failure given by:

$$P_{\text{failure}} = P(\text{I}) + P(\text{II}) - P(\text{I} \cap \text{II})$$

where $P(\text{I})$ is the conditional probability of sufficient rain to trigger a failure given the toe is removed

$P(\text{II})$ is the conditional probability of excessive rain to trigger a failure given that the toe is still in place

$P(\text{I} \cap \text{II})$ is the joint probability of $P(\text{I})$ occurring at the same time as $P(\text{II})$.

We can assume that the events causing extreme coastal storms and rainfall are independent, although in reality there may be some positive correlation between a storm and rainfall during lesser events.

$P_{total} = P\,(\text{I}) + P\,(\text{II})$

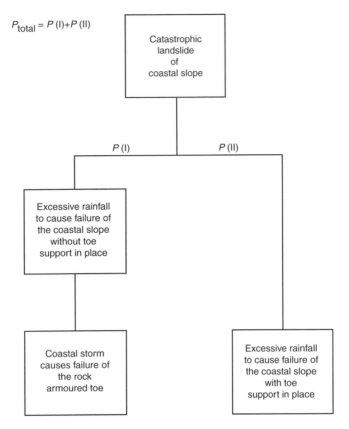

Figure C2.6. Fault tree leading to a catastrophic failure of the coastal slope

If the assumption is made that the joint probability event is negligible then the equation is rewritten as:

$$P_{failure} = P(\text{I}) + P(\text{II})$$

The prediction of $P(\text{I})$ is based on the probability of the toe moving and the slope then failing due to rainfall. The rock revetment is designed to withstand a 1:100 year event and, without further information, this can be assumed to be the trigger annual probability, which is 0·01 (i.e. 1/100). The annual probability of failure of the coastal slope along the whole Undercliff of which Castle Cove lies within has been investigated extensively by a number of experienced engineers (Lee *et al.*, 1998; Hall *et al.*, 2000). (For the benefit of this case study analysis, it has been assumed that probabilities predicted for the Undercliff are representative of Castle Cove, which is strictly an inaccurate assumption.) A number of conceptual landslide models have been developed that have allowed the annual probability of failure to be predicted for the slope before the project was constructed. Historical analysis was used to predict the trigger levels for failure, given a defined rainfall, which could then be linked to a probability of occurrence. This resulted in a prediction of the annual probability of rainfall and the slope failing to be 0·06 (Table C2.8). This prediction compares well with the benefits assessment made by the designer who stated that landslides would start to occur within 10 to 20 years if a 'do nothing' approach was adopted.

The annual probability of failure due to the toe support being lost ($P(I)$) is, therefore, provided by:

$$P(I) = 0.01 \times 0.06 = 0.0006$$

Once the project is constructed, the conditional probabilities of the slope failing due to rainfall will decrease due to improved drainage (i.e. designed to withstand a return period of approximately 1 in 100 years) and increased toe support. Therefore, the original prediction of the annual conditional probability of rainfall and slope failure can be adjusted to 0·001 (Table C2.9).

Table C2.8. The annual probability of failure with the toe support not in place

Rainfall intensity	Annual probability of rainfall occurring	Annual conditional probability of failure if rainfall occurs	Annual probability of rainfall and slope failure
<410	0·80	0·0	0·000
410–540	0·16	0·2	0·032
540–640	0·03	0·6	0·018
>640	0·01	1·0	0·010
		Annual probability	0·060

Table C2.9.   The annual probability of failure with the toe support in place

Rainfall intensity	Annual probability of rainfall occurring	Annual conditional probability of failure if rainfall occurs	Annual probability of rainfall and slope failure
<410	0·80	0·0	0·000
410–540	0·16	0·0	0·000
540–640	0·03	0·0	0·000
>640	0·01	0·1	0·001
		Annual probability	0·001

The annual probability of failure due to the toe support remaining, but there being excessive rainfall ($P(\text{II})$), is, therefore, provided by multiplying the probability of the toe still being in place by the annual probability of failure, which gives the following answer:

$$P(\text{II}) = (1 - 0.01) \times 0.001 = 0.00099$$

If $P(\text{I})$ and $P(\text{II})$ are inserted into the original equation, the annual probability of failure for the coastal slope is:

$$P_{failure} = 0.0006 + 0.00099 = 0.00159 \text{ (i.e. of the order of } 10^{-3})$$

This annual probability of failure, therefore, has a return period of approximately 600 years, which gives the scheme an exceedance probability of occurrence of less than 0·1, assuming a design life of 50 years. The designer now has an understanding of the scheme's probability of failure by considering the magnitude of hydraulic design conditions and the failure mechanisms of the slope. This type of assessment is very difficult to do with any certainty and can only be properly achieved by using an engineer with sufficient experience and competence.

The value of 10^{-3} seems a reasonable risk level given that structural codes of practice are largely based on a probability of failure of 10^{-4}, which, in general, were produced for structures with a longer design life of say 100 years. The latter gives a probability of

Table C2.10. Risk comparison between 'with project' and 'do nothing' strategy (continued overleaf)

Year	Discount factor	With project				'Do nothing' strategy			
		P_{occur}	$P_{does\ not}$	Damage: £	Present value: £	P_{occur}	$P_{does\ not}$	Damage: £	Present value: £
0	1·00	0·00159	0·99841	3 450 000	5486	0·06000	0·94000	3 450 000	207 000
1	0·94	0·00159	0·99682	3 450 000	5148	0·05640	0·88360	3 450 000	182 905
2	0·89	0·00158	0·99524	3 450 000	4867	0·05302	0·83058	3 450 000	162 786
3	0·84	0·00158	0·99366	3 450 000	4586	0·04984	0·78075	3 450 000	144 422
4	0·79	0·00158	0·99208	3 450 000	4306	0·04684	0·73390	3 450 000	127 676
5	0·75	0·00158	0·99050	3 450 000	4082	0·04403	0·68987	3 450 000	113 939
6	0·70	0·00157	0·98892	3 450 000	3803	0·04139	0·64848	3 450 000	99 962
7	0·67	0·00157	0·98735	3 450 000	3635	0·03891	0·60957	3 450 000	89 937
8	0·63	0·00157	0·98578	3 450 000	3412	0·03657	0·57299	3 450 000	79 494
9	0·59	0·00157	0·98421	3 450 000	3190	0·03438	0·53862	3 450 000	69 980
10	0·56	0·00156	0·98265	3 450 000	3023	0·03232	0·50630	3 450 000	62 436
11	0·53	0·00156	0·98109	3 450 000	2857	0·03038	0·47592	3 450 000	55 546
12	0·50	0·00156	0·97953	3 450 000	2691	0·02856	0·44737	3 450 000	49 258
13	0·47	0·00156	0·97797	3 450 000	2525	0·02684	0·42052	3 450 000	43 524
14	0·44	0·00155	0·97641	3 450 000	2360	0·02523	0·39529	3 450 000	38 301
15	0·42	0·00155	0·97486	3 450 000	2250	0·02372	0·37157	3 450 000	34 367

16	0·39	0·00155	0·97331	3 450 000	2086	0·02229	0·34928	3 450 000	29 997
17	0·37	0·00155	0·97176	3 450 000	1975	0·02096	0·32832	3 450 000	26 751
18	0·35	0·00155	0·97022	3 450 000	1866	0·01970	0·30862	3 450 000	23 787
19	0·33	0·00154	0·96868	3 450 000	1756	0·01852	0·29011	3 450 000	21 082
20	0·31	0·00154	0·96714	3 450 000	1647	0·01741	0·27270	3 450 000	18 616
21	0·29	0·00154	0·96560	3 450 000	1539	0·01636	0·25634	3 450 000	16 370
22	0·28	0·00154	0·96406	3 450 000	1483	0·01538	0·24096	3 450 000	14 857
23	0·26	0·00153	0·96253	3 450 000	1375	0·01446	0·22650	3 450 000	12 968
24	0·25	0·00153	0·96100	3 450 000	1320	0·01359	0·21291	3 450 000	11 721
25	0·23	0·00153	0·95947	3 450 000	1212	0·01277	0·20014	3 450 000	10 137
26	0·22	0·00153	0·95795	3 450 000	1158	0·01201	0·18813	3 450 000	9114
27	0·21	0·00152	0·95642	3 450 000	1104	0·01129	0·17684	3 450 000	8178
28	0·20	0·00152	0·95490	3 450 000	1049	0·01061	0·16623	3 450 000	7321
29	0·18	0·00152	0·95338	3 450 000	943	0·00997	0·15626	3 450 000	6194
30	0·17	0·00152	0·95187	3 450 000	889	0·00938	0·14688	3 450 000	5499
31	0·16	0·00151	0·95035	3 450 000	835	0·00881	0·13807	3 450 000	4865
32	0·15	0·00151	0·94884	3 450 000	782	0·00828	0·12978	3 450 000	4287
33	0·15	0·00151	0·94733	3 450 000	781	0·00779	0·12200	3 450 000	4030
34	0·14	0·00151	0·94583	3 450 000	728	0·00732	0·11468	3 450 000	3535
35	0·13	0·00150	0·94432	3 450 000	674	0·00688	0·10780	3 450 000	3086

Year	Discount factor	With project				'Do nothing' strategy			
		P_{occur}	$P_{does\ not}$	Damage: £	Present value: £	P_{occur}	$P_{does\ not}$	Damage: £	Present value: £
36	0·12	0·00150	0·94282	3 450 000	622	0·00647	0·10133	3 450 000	2678
37	0·12	0·00150	0·94132	3 450 000	621	0·00608	0·09525	3 450 000	2517
38	0·11	0·00150	0·93983	3 450 000	568	0·00571	0·08953	3 450 000	2169
39	0·10	0·00149	0·93833	3 450 000	516	0·00537	0·08416	3 450 000	1853
40	0·10	0·00149	0·93684	3 450 000	515	0·00505	0·07911	3 450 000	1742
41	0·09	0·00149	0·93535	3 450 000	463	0·00475	0·07437	3 450 000	1474
42	0·09	0·00149	0·93386	3 450 000	462	0·00446	0·06990	3 450 000	1385
43	0·08	0·00148	0·93238	3 450 000	410	0·00419	0·06571	3 450 000	1158
44	0·08	0·00148	0·93090	3 450 000	409	0·00394	0·06177	3 450 000	1088
45	0·07	0·00148	0·92942	3 450 000	357	0·00371	0·05806	3 450 000	895
46	0·07	0·00148	0·92794	3 450 000	357	0·00348	0·05458	3 450 000	841
47	0·06	0·00148	0·92646	3 450 000	305	0·00327	0·05130	3 450 000	678
48	0·06	0·00147	0·92499	3 450 000	305	0·00308	0·04822	3 450 000	637
49	0·06	0·00147	0·92352	3 450 000	304	0·00289	0·04533	3 450 000	599
					89 636				1 823 643

exceedance of less than 0·001, which is very unlikely to occur during the lifetime of the structure. The decision that the designer must make is whether or not it is worth making the design safer. This could have been achieved by the flattening of the slope further, installing more efficient drainage, removing the clay layer and constructing a more effective drainage surface, etc.

The benefit of linking a design to a probability of failure is that checks can be made such that the residual risk left after the scheme is acceptable to the client and the public as a whole. In the original design, the designer made the assumption that the residual risk was considered to be minimal once the new scheme was constructed. However, the residual risk in some designs may be significant and may need to be calculated as part of the risk equation. The true risk can be quantified by considering the present value over the scheme's design life. The probability of the slope failing in a given year is determined by multiplying the annual probability of failure by the probability that the slope has not failed in the previous year. The risk (i.e. the probability that slope failure occurs multiplied by the expected damage) is then multiplied by the discount factor (6%) for that particular year. The summation of all the discounted risk values is the present value of risk for the design life.

Using this approach, the actual risk of failure with the scheme in place was predicted to be £90 000 compared with £1·8 million without the scheme in place (Table C2.10). Thus, the exclusion of the residual losses in this particular design had minimal impact on the BCR, with a small negative variance of 0·06 from 1·23 to 1·17, assuming that the capital cost of the scheme was approximately £1·4 million. In reflection, it is reasonable that the designer assumed that the residual losses were negligible and not worth assessing because reasonable design events had been selected with a high factor of safety as well. It is noteworthy to state that residual risk assessments are now required as part of DEFRA procedures for flood and coastal protection schemes.

In summary, the deterministic slope analysis method was an effective method, which allowed the designer to visualise quite effectively the different elements and to select the appropriate design tools. A probabilistic approach in this case study was useful in the initial risk analysis in order to gain a broad understanding of the risks involved and the key design parameters that should be considered in the deterministic design.

C2.7. CONCLUSIONS

The conclusions of the case study are as follows.

- Considerable research has been undertaken at the site and, therefore, the adoption of a scheme was only agreed once the slope had started to fail rather than acting before failure occured. The time in which to submit grant application was lengthy and, as a result, perhaps the assets within the coastal slope were at a higher risk level for a longer time than was necessary.
- The funders had an influence on setting budgets for the design based on benefit assessments, but in the end it was the influence of the Nature Conservation Officer with the installation of a Gault clay layer who had the most influence on the final outcome of the design.
- The provision of a toe support with the rock revetment was a pivotal part of the design which required interaction between the hydraulic and geotechnical disciplines to produce an effective design, both used a deterministic approach to design.
- The acceptable risk level was not based on optimising risk but rather on eliminating risk by designing a structure that had a very small probability of failure. This was achieved by selecting appropriate design events for the design life required.

C2.8. ACKNOWLEDGEMENTS

HR Wallingford would like to thank High Point Rendel and the Isle of Wight. Special thanks must go to Mr Steve Fort and Mr Hugo Wood, who were the principle contacts at High Point Rendel.

C2.9. REFERENCES

British Standards Institution (BSI) (2000). *Code of practice for maritime structures — Part 1.General criteria.* BSI, London, BS6349-1 .

British Standards Institution (BSI) (1983). *Code of practice for earthworks.* BSI, London, BS 6031.

Hall J. W., Lee E. M. and Meadowcroft I. C. (2000). Risk-based benefit assessment of coastal cliff protection. *Proceedings of the Institution of Civil Engineers, Water and Maritime Engineering*, **142**, No. 3. ISSN 0965 0946.

Lee, E. M., Moore, R. and McInnes, R. (1998). Assessment of the probability of landslide reactivation: Isle of Wight Undercliff UK. *Proceedings of the 8th International Association of Engineering Geology and the Environment Congress, Vancouver*, pp. 1315–1321.

Ministry of Agriculture, Fisheries and Food (1993). *Flood and coastal defence project appraisal guidance notes*. MAFF, London.

Clark, A. and Fort, D. S. (1998). Castle Cove landslide stabilisation and coast protection, Ventnor, Isle of Wight, UK. *Proccedings of the 8th International Association of Engineering Geology and the Environment Congress, Vancouver*.

Case study 3.
Dawlish seawall

C3.1. OBJECTIVES OF THE CASE STUDY

The objectives of this case study are to:

- describe the design process
- demonstrate the principle of setting risk levels in terms of its own individual level of tolerability (i.e. unacceptable, tolerable and broadly acceptable).

C3.2. BACKGROUND

C3.2.1. Description of the site and the seawall

Dawlish sea defence is a coastal defence that provides direct protection to a twin track railway line and the town of Dawlish and Teignmouth. The railway line is the only rail link from west Devon and Cornwall to London. The seawall itself extends from the mouth of the Exmouth to Teignmouth, a length of 6·1 km (Figure C3.1).

The current structure is a concave seawall with a low return wall and toe protection (Figure C3.2). The materials used to construct the wall are limestone blocks that have a minimum thickness of 650 mm, with the infill material behind the wall to support the tracks consisting of a permeable mixture of rock and shingle. The seawall itself is founded on soft breccia sandstone. A small beach provides some protection to the toe of the structure.

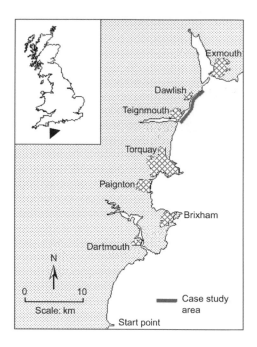

Figure C3.1.   Dawlish sea defence location map

C3.2.2. Development of the seawall at Dawlish

Construction of the existing Dawlish seawall commenced in 1845 under the direction of Brunel as part of the South Devon Railway. The management and maintenance of the seawall at Dawlish has been an engineering challenge right through its lifetime. Even during the construction phase, considerable problems were experienced in 1846 with a breach in the wall. Various remedial and reconstruction works were carried out over the early years to mitigate against a number of problems including sections of the wall collapsing and the railway track being washed away. Most of the wall comprised the original construction, although a few sections have been completely rebuilt at one time or another.

In 1997, after a century and a half of reactive repairs, Railtrack (previously British Rail) commenced work on providing a large scale

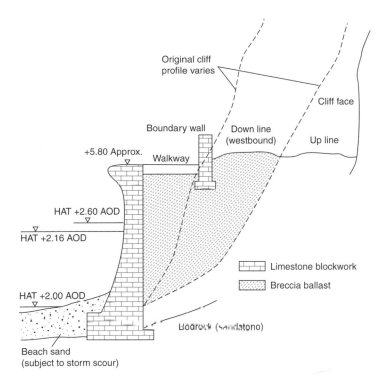

Figure C3.2   Typical section through original seawall

concrete toe to sections of the wall perceived to be most at risk. As
this appeared to be having some success in withstanding wave and
tidal action, it was decided to look at other possible cost-effective
solutions to maintaining the wall and to confirm that the present
method of toe protection was a preferred solution. Hyder Consulting
were commissioned to undertake a feasibility study with the aim of
reducing the overall maintenance bill of the seawall and to reduce
annual penalty payments for speed restrictions and closures to the
train-operating companies that use the line. Before the feasibility
study, Railtrack were spending in excess of £1 million annually on
the seawall, comprising minor repairs, maintenance and major works.
The aim of the feasibility study was to formulate a strategy to achieve
the project objectives over the next 20–25 years (Hyder Consulting,
1998).

The main problems that were identified by the feasibility study were as follows.

Reduction in beach levels

Beach levels vary considerable along the wall, exposing the toe of the structure. Control of the beach is aided by a groyne system, which has become ineffective at certain locations.

Undermining of the wall

Exposure of the toe of the structure has led to the infill material behind the wall becoming washed out, leaving a number of voids behind the structure.

Voids behind the seawall

The large voids that have been created by action of groundwater and waves removes the structural support to the railway track and the walkway. Large grouting operations in the past involved the use typically of up to 150 t of material. On occasions, the grout has been seen to come out of the seawall further along its length, demonstrating the complexity of the voids right along the seawall.

Overtopping

With low beach levels, the seawall is exposed to large waves that overtop the wall and cause the railway line to be closed due either to waves hitting the trains or the risk of the track support being damaged. In lesser storms, single track working is brought into operation at reduced speeds, although this also has an impact on train delays.

Study conclusions

Confirmation was provided from the feasibility study that the initiative to construct a new concrete toe to reduce the undermining and loosening of fill material supporting the track gave the highest cost benefit. This was to be coupled with improved drainage through the wall to reduce groundwater causing instabilities in the fill material. Design of this toe was checked and found to be sufficiently dimensioned to provide a reasonable life expectancy.

Other recommendations ranged from a new shape of capping to the wall to reduce the overtopping, to the introduction of a revised signalling arrangement to provide more efficient train path operations during periods of single line working. While not providing such robust business cases, these topics will be looked at when the time is appropriate.

Further cost reductions have resulted from recommendations on methods of procurement, including obtaining competitive lump sums for larger sections of the work, including the toe reconstruction. This work had previously been carried out on a reimbursable basis.

The present situation

Where there are significant voids in the track support structure, attempts are made to fill these voids. The toe detail which cannot be constructed in one phase, is prioritised on the basis of where it is most needed. Some localised parts of the frontage have also undergone re-facing of the existing seawall. No work on the beach has been undertaken at this time, nor any attempt to reduce overtopping of the seawall.

C3.3.   INFLUENCE OF STAKEHOLDERS ON THE DESIGN PROCESS

The responsibility of the seawall lies with the operator of the railway line under the South Devon Railways Act 1844, which is now Railtrack.

The responsibilities of the different parties involved are summarised in Table C3.1.

As part of the consultation process, all stakeholders have stated what they would accept, tolerate and not accept regarding any work undertaken along the seawall. The reactions of the consulted bodies are summarised in terms of boundaries of acceptable risk in Table C3.2.

From Table C3.2, it is clear that Railtrack is the only stakeholder who will gain or lose directly in a financial sense. The rest of the parties seem reasonably happy to tolerate no change to the current situation and would accept any enhancement that came along. As a private company, Railtrack found it difficult to obtain any financial support from other stakeholders even though enhancement might well have improved tourism to the area or continued maintenance of the seawall would have minimised coastal erosion. As a consequence, the final design was very much focused on Railtrack's needs. If external funding had come forward then the final design may have been significantly different but would still have met the

Table C3.1.   Organisations that have a responsibility or an influence on the seawall

Parties	Responsibility	Responsibility in Dawlish sea defence
Railtrack	To run and maintain railway lines	To run the railway line on top of Dawlish seawall and to maintain the seawall
Countryside Commission	To protect the countryside of England and Wales, and is the designatory body for Areas of Outstanding Natural Beauty	To award grant aid to local authorities for upkeep and improvements to the South West Coast Path
South Hams Coast and Countryside Service	Set up in 1997 to manage the coast path as the agent of the local highway authority	To comment on the proposal and to arrange Temporary Highway Closure Notices if necessary
English Nature	Any construction work in an area designed as an SSSI	The rear of the Dawlish Beach and Coryton's Cove is an SSSI
DTI	To work in tidal water requires a consent of the Secretary of State	Give licences for any work carried out below mean sea-level which may effect navigation
DEFRA	Issuing licences for offshore dredging and dumping	Issue licences if required
	Also, responsible for coast protection	Also, provide grant aid for coast protection only if justified
Dawlish Town Council (DTC)	Teignbridge District Council (TDC) handles all applications for planning permission	DTC has regular planning meeting with TDC where any items directly related to Dawlish are discussed
Teignbridge District Council	Responsible for coastal protection and planning	Only responsible for planning issues
Environment Agency	The Environment Agency and the drainage authority exercise a supervisory role over all matters relating to flood defence	No responsibility, as the area under consideration is coastal protection, principally the local authority (TDC) would control planning matter
Crown Estate	Ownership of foreshore	Ambiguous, an agreement with Great Western Railways in 1909 allowed the construction of groynes

Table C3.2. Acceptable risk levels set out by different parties

Party	Unacceptable	Tolerable	Broadly acceptable
Railtrack	NPV/outlay <= 1 No financial benefit	NPV/outlay > 1 Financial benefit to Railtrack	NPV/outlay > 1 Improved financial benefit to Railtrack
Countryside Commission	Damage to coast path	Keep status quo	Enhancement
South Hams Coast and Countryside Service	Damage to coast path	Keep status quo	Enhancement
English Nature	Construction on or damage to SSSI	Keep status quo	Enhancement
DTI	Negative impact on navigation	Mitigation against negative impacts	Keep status quo
DEFRA	Negative environmental impact on fisheries	Would provide grant aid only if a significant negative impact to the national economy	Keep status quo
Dawlish Town Council	Negative impact on tourism	Keep status quo	Enhancement
Teignbridge District Council	Negative impact on tourism	Keep status quo	Enhancement
Environment Agency	Increase risk of flooding	Keep status quo	Keep status quo
Crown Estate	Negative impact on the beach	Keep status quo	Keep status quo

objectives of Railtrack and also made improvements to the local area. An example of an external funder that might have become involved is DEFRA, who is responsible for coast protection in England and Wales. However, funding from this source is usually restricted to grants to local authorities or the regional administration of the Environment Agency and are not made direct to individuals or companies as it is felt that they should not benefit from taxpayers' money.

C3.4. DESIGN PROCESS

C3.4.1. Identification of a need

As discussed Section C3.2.2, Dawlish seawall has a long history of maintenance problems stretching back to its conception. With the privatisation of the rail infrastructure and the formation of Railtrack, it became important to focus on solutions to these problems that provided value to their business and, most importantly, to their shareholders.

C3.4.2. Functional analysis

A value management workshop was held in February 1998 for all key staff involved with the management of the seawall to define the functionality of the structure. The functions that were defined by the workshop were:

- to reduce annual maintenance costs
- to improve services by reducing the number of closures and speed restrictions
- a design life of 50 years.

C.3.4.3. Generate alternative solutions

During the same workshop, 92 proposals were generated during a brainstorming session. These proposals were grouped into the following areas:

- changing beach levels
- offshore structures
- changes to wall structure
- contractual aspects
- permanent way/signalling improvements
- drainage
- cliff works
- obtain alternative funding (EEC funding/lottery money)

- monitoring studies
- construction issues/planning
- operational issues.

C3.4.4. Comparison and selection

The proposals in the brainstorming session were then screened and rated by the participants of the workshop using the following marking system:

1 – good idea — develop further (68 proposals)
2 = good idea — but of little value or difficult to implement (12 proposals)
3 = wacky/unachievable (8 proposals).

All of the proposals that satisfied a scoring of one were then reviewed by Hyder Consulting and re-rated based on the following criteria:

Grade A = likely to achieve objective
Grade B = may achieve objective but feasible/little benefit
Grade C = discard.

An engineering value meeting was then held to discuss which proposals would be carried forward to a more detailed level of design. Several ideas were considered further but then discarded; some of these were as follows.

- After a probabilistic analysis of waves and water levels, it is concluded that green water overtopping within the next 50 years is very unlikely, and so was decided that a new coping for the seawall is not necessary. Wave-induced overtopping happens frequently, but Railtrack felt that at present the inconvenience was tolerable compared to the risk of a catastrophic failure of the track support.
- It is concluded that raising groynes by 500 mm would have a beneficial impact on the shape of the beach, but there is fear that the forces on the groynes will be such that they will not last as long as the existing groynes. Suggestions were made to construct a test section and monitor their performance, although at this time the work has not yet commenced.
- Offshore structures and a beach dewatering system were found to be not viable or economic. To raise the beach level, beach

nourishment has been carried out in the past, making use of dredging from Teignmouth docks. This dredging has since stopped and beach nourishment is now thought to be not economic.

The selection process, therefore, put forward the following options (as described above) to be designed in detail and implemented:

- detection of voids and grouting them
- construction of a new stepped toe
- new facing works at Dawlish train station
- masonry repairs to the face of the wall to sections that had been significantly damaged
- concrete spraying at the toe of the wall as an emergency measure.

C3.4.5. Detailed design

The detailed design was based largely on intuitive engineering judgement, rather than detailed modelling or design codes. The designers depended largely on looking at the performance of the materials and structures already in-situ and making a decision based on observation. The main design issues for each scheme are summarised as follows.

- *Detection of voids and grouting them.* Two void detection systems (radar and sonar) were tested with radar being the preferred method. Grouting techniques were used that had been tested previously at different areas of the site.
- *Construction of a new stepped toe.* The toe was designed by the contractor to be bulky enough to remain in place and durable enough to resist high levels of abrasion over the 50-year design life. Design decisions were made largely on the performance of similar material along the frontage. To reduce uncertainty in the performance of the design, a test section was constructed and monitored, before a wide-scale programme was commissioned. As mentioned previously, a small amendment to the shape was recommended for optimum life expectancy.
- *New facing works at Dawlish train station.* The designers followed a similar design as the previous one in-situ, but made some enhancements by considering research undertaken by HR Wallingford on the shapes of seawalls.

- *Masonry repairs to the face of the wall.* Rather than a reactive maintenance plan of only doing works when damaged, panels were selected and prioritised by the local Railtrack engineer.
- *Concrete spraying at the toe of the wall as an emergency measure.* This had been the preferred method, as developed in the late 1980s, and was only used when emergency works were required. In practice, the life expectancy of this method proved to be very short and this encouraged the introduction of a new toe detail.

C3.4.6. Construction

The construction of the design was based on prioritisation made by the engineer responsible for the inspection and operation of trains along the seawall. The contract was set up so that the contractor was given time to carry out the works but with flexibility to do so when it best suited him. This reduced the problem of weather delay and other construction risks.

C3.4.7. Management

The seawall and the track are inspected on a regular basis and, therefore, problems can be located at an early stage.

C3.4.8. Decommission

The structure has been at Dawlish for a long time and its role as a railway line means it is unlikely that the structure itself will be decommissioned but rather it will be managed and maintained indefinitely. The only major change will come if Railtrack decide to abandon the railway line or something else occurs of a similar magnitude.

C3.5. ACCEPTABLE RISK ISSUES

This section of the case study was undertaken solely by HR Wallingford as part of the research project. Analysis has been used to demonstrate the basic principles set out in this book and should not be considered as accurate design data for other projects.

C3.5.1. Acceptable Risk Bubble

Conceptually, the Acceptable Risk Bubble is a powerful tool to help the designer manage multiple risks efficiently. In practice, this thought process might be used by a number of designers already but there is very little information available on the key steps required to actually manage a multi-attribute problem. This case study has been used to illustrate how such a process could work in practice and demonstrates the benefit of undertaking the analysis.

The main steps that are considered are:

- identify stakeholders and risk owners
- brainstorm risks/concerns and rank them
- select top level concerns
- define value curves for each concern
- define broadly acceptable, mid-tolerable and unacceptable levels
- score projects against Acceptable Risk Bubble.

The possible process is now discussed below.

Identify stakeholders
Stakeholders identified were:

- Railtrack
- Teignbridge District Council
- Dawlish Town Council
- Countryside Commission
- South Hams Coast and Countryside Service.

The key risk owners were:

- Railtrack (funder and designer)
- Teignbridge District Council (local authority).

Brainstorm risks/concerns, rank them and select top level risks/concerns
The following risks/concerns were highlighted as being key to the success of the project:

- engineering
- economics
- operation of the railway

- coast protection
- flooding
- tourism.

Define broadly acceptable, mid-tolerable and unacceptable levels
Table C3.3 defines qualitatively the value curve for each risk/concern.

Score projects against Acceptable Risk Bubble
The criteria for scoring projects were based on each of the risk categories as:

Table C3.3. Dawlish — acceptable risk matrix

Risk/concern	Broadly acceptable	Mid-tolerable	Unacceptable
Engineering	Complete repair of the seawall	Systematic piecemeal approach to repairs, prioritise on high risk sections	Patch and repair only when section becomes critical
Economics	High BCR	BCR equal to 1	Gain no economic benefit from undertaking works
Operation	No delays due to wave-induced overtopping or breaching of the seawall	Infrequent delays due to wave-induced overtopping	Large delays due to seawall breaching causing loss of track support material
Coast protection	No further erosion	Status quo	Large losses of land or property put into danger
Flooding	Overtopping spray allowed up to 200-year sea condition	Status quo	Frequent flooding
Tourism	Enhancement to the tourism industry	Status quo	Negative impact on tourism, loss of beach and coast path

- broadly acceptable equal to or less than 1
- tolerable between 1 and 3 (mid-point of tolerability equals 2)
- unacceptable equal to or greater than 3.

In this case study, a comparison was made between the selected project and a project that was rejected on economic grounds. The scoring estimates for both projects are given in Tables C3.4 and C3.5 and the results plotted onto the Acceptable Risk Bubble (Figure C3.3). The analysis shows that improving the beach seems a stronger option when comparing against the majority of risks but the economics mean that the overall project is unacceptable. Further work on estimating construction costs and benefits or a change in the risk levels on the economic axis may have made the option more acceptable.

C3.5.2. Distribution of risk levels

Designing a structural system so that there are no obvious weak links is a clear goal of any designer, if there is a belief in the analogy that the chain is only as strong as its weakest link. In the vast amount of

Table C3.4. Dawlish — acceptable risk scoring table {Project (Seawall)}

Risk/concern	Score	Reason
Engineering	1·8	Systematic piecemeal approach promoting active management of the seawall rather than reactive management
Economics	1·5	Reasonable return on Railtrack's investment
Operation	2·2	Apart from reducing the risk of breaching, there are no improvements to operational safety with regard to wave-induced overtopping
Coast protection	1·5	Reduced the risk of breaching occurring through undermining of toe by coastal processes, but some residual risk remaining with the flow of groundwater through the structure
Flooding	2·0	No positive impact on reducing flooding
Tourism	2·0	No impact on tourism

Table C3.5.   Dawlish — acceptable risk scoring table {Project (Beach)}

Risk/concern	Score	Reason
Engineering	1·0	Produces a solution for the whole length of the seawall
Economics	5·0	Poor economic return of Railtrack's investment
Operation	1·0	Reduce wave induced overtopping decreasing operation downtime through bad weather
Coast protection	1·5	Seawall protected by beach
Flooding	1·5	Positive impact in reducing flooding
Tourism	1·5	Improved beach amenity to the area

cases in coastal and fluvial engineering, the designer is refurbishing an existing structure. Where the existing structure is ascertained to be a weak link, it may be better to keep that existing structure and instead provide more protection to the existing structure (e.g. raise the beach levels in front of an existing seawall). There is, however, a clear need to gain a broader understanding of what designers are currently doing

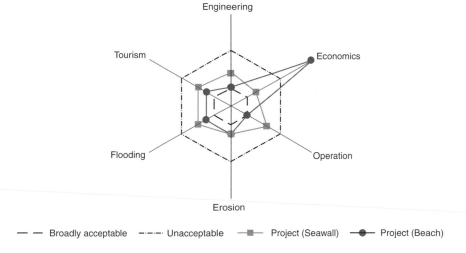

Figure C3.3.   Dawlish — Acceptable Risk Bubble

and where they can focus efforts to improve the efficiency of their designs. The case studies have been used to illustrate the risk levels accepted for each structural element and how these risk levels potentially interact as part of the overall system.

The main steps in this analysis of the case study analysis were to:

- identify key structural elements of the system
- identify the design discipline they are affiliated to
- identify their reliance on other structural elements
- define the lifetime probability of failure accepted for each element.

The seawall at Dawlish can be subdivided into two main structural elements, as follows.

- *Beach*. The beach has remained stable over a long period and as a result has required historically very little maintenance. For the purposes of this analysis, a lifetime probability of 0·05 is assumed, which equates to a 1000-year return period over a working design life of 50 years. The beach can be described as a hydraulic structure but relies on no support from the other structural elements in order to maintain its own integrity.
- *Seawall structure*. The seawall structure itself was constructed over 100 years ago and in recent times has required high maintenance to keep the seawall structurally sound. The solution adopted by the owners was to construct a new toe detail to minimise undermining and to repair the masonry blockwork to maintain the face. The structure has not been designed to be stable against a specific loading event, as the wall has remained largely stable over a long period of time. The design criterion used was to use materials that have a fatigue life of at least 20 to 30 years in a coastal environment. For the purpose of this analysis, the assumption of lifetime probability for this element has been estimated as 0·2, which equates to a 200-year return period over a working design life of 50 years. This assumption is based on a crude assessment of design criteria used for other seawall designs. The seawall structure is focused on structural design, but its success is reliant on the beach staying in place in order to maintain its own integrity.
- *Seawall crest*. Wave-induced overtopping of the crest causes mainly disruption to the operation of the railway line rather than

structural damage. It was decided not to actively reduce overtopping with this solution. The accepted risk levels are predicted from historical records of overtopping causing disruption. The design of the crest is part of the hydraulic discipline but it still relies on the stability of the seawall structure and the beach to maintain similar risk levels currently observed on site.

- *Track support.* Due to undermining of the toe, track support material was washed out under the seawall either by wave action or groundwater movement. Risk of the track support failing was managed through the construction of the new toe detail and the accepted risk level of the structural element was assumed to be the same as that observed historically. The track support material is part of the geotechnical discipline but relies on the stability of the seawall structure and the beach to maintain its own integrity.

- *Coastal slope.* The same argument applies here as for track support, the risk being managed through the new toe detail and therefore the risk level is assumed to be equivalent of historical observations. This element sits within the geotechnical discipline and relies on the stability of the seawall structure and the beach to maintain its own integrity.

Table C3.6. Dawlish — accepted risk levels for each structural element

Structural element	Discipline*	Design return period: years	Working design life: years	Individual lifetime probability of failure: %†
Beach	H	1000	50	5
Seawall structure	S	200	50	22
Seawall crest	H	7	50	99
Track support	G	1	3	100
Coastal slope	G	1	3	100

* H = Hydraulics; G = Geotechnics; and S = Structures.

†$P_{individual} = 1 - (1 - 1/TR)^n$ where TR is the design return period and n is the design life.

The crude analysis of accepted risk levels in terms of quantifying the lifetime probability of failure over the design working life shows a wide range of extreme risk levels accepted for each individual element (Table C3.6). When the elements are considered as part of an overall system, a large reduction in the systems risk level is observed by reducing the risk of undermining through the construction of a new toe detail (Table C3.7). In this particular case study, the primary design discipline controlling the design seems to be structural, but in fact the structure itself has been designed to withstand the long-term abrasion of the environment which is largely driven by the hydraulic discipline. It is interesting how one element can have such a large impact on controlling the overall risk level within the whole system,

Table C3.7. Dawlish — accepted risk levels for each structural element as part of a system

Structural element	Reliability function	Conditional lifetime probability of failure: %*	Reduction in individual conditional lifetime probability: %†
Beach	Beach (P_1)	1	0
Seawall structure	Beach (P_1) Sewall structure (P_2)	1	22
Sewall crest	Beach remaining (P_1) Seawall structure (P_2) Sewall crest (P_3)	21	78
Track support	Beach (P_1) Seawall structure (P_2) Track support (P_3)	1	99
Coastal slope	Beach (P_1) Sewall structure (P_2) Coastal slope(P_3)	1	99

*$P_{individual} = 1 - \left(1 - 1/TR\right)^n$ where TR is the design return period and n is the design life.

†$P_{total} = P_1 \times P_2 \times \ldots$ and so on.

while some of the elements on their own are still reasonably weak. This point illustrates the value of protecting elements rather than trying to strengthen them.

C3.5.3. Dealing with different levels of tolerability in design

The risk that Railtrack is trying to minimise is the delay of trains because Railtrack is liable to pay a penalty for every minute a train is delayed due to a track closure or a speed restriction. Any properties, assets and lives protected by the seawall are of no (economic) interest to Railtrack.

The risk of delay to trains can be initiated through a number of different events impacting on the structure. The main elements of the seawall comprise the beach, the seawall itself, the fill material supporting the track and a cliff face (in localised stretches), which have to interaction with these events (Figure C3.4).

The interactions of these events are best summarised in a fault tree that shows how a number of different events can lead to a delay of the trains (Figure C3.5). The three main faults are as follows.

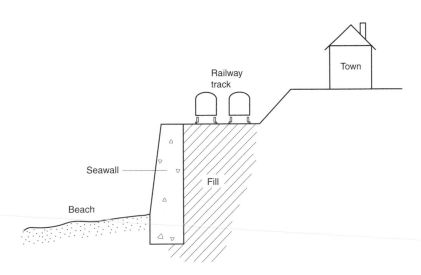

Figure C3.4.   Description of the sea defence system

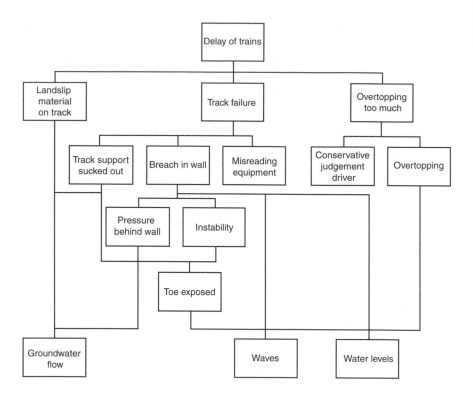

Figure C3.5. Fault tree leading to a train delay

- *A failure of the track* — the main failures of the track can be caused by either a breach in the seawall through exposure to varying waves and water levels or by loss of track support material through washing out of material under the seawall due either to wave action or groundwater movement. In some cases, closure or speed restriction may be due to misreading of monitoring equipment rather than an actual structural failure.
- *Excessive overtopping of the seawall* — the track is closed due to overtopping if the Operations Manager decides that it is unsafe to drive across the seawall or if weather forecasts exceed a certain storm trigger level.
- *A landslide of material onto the track* — occasionally, some landslide material falls on the track and causes some delay of trains.

An analysis of events that caused disruption during 1995 and 1996 was undertaken as part of the feasibility study. Using the figures published in the feasibility study, an attempt was made during this research to quantify the levels of risk for each failure event to illustrate why Railtrack made certain decisions in setting out the needs of the project. The main assumptions in this analysis were:

- on average, there are three trains per hour every hour passing on the downline and three per hour on the upline, based on the current timetable
- a train takes approximately three minutes to cross the 6·1 km section of the seawall under normal operating conditions (average speed of 80 mph)
- the number of minutes delay due to a *track closure* is the product of the number of trains that have to wait and the average time they have to wait
- the number of minutes delay due to a *speed restriction* is the product of the number of trains that arrive during the speed restriction and the time they need to cross the 6·1 km section minus three minutes.

The results of this crude analysis showed that failure of the track support was the main area of risk, with either a catastrophic failure of the seawall or voids forming under the track causing deflection problems (Table C3.8). Even though overtopping is the most frequent event, the typical consequences of an overtopping event (200 minutes) are much smaller, as the track can still operate with a speed restriction in force, compared to complete closure arising from a catastrophic failure (1 million minutes). The financial consequence depends on the nature of the speed restriction, the type of train operating on the line and a number of other factors that are bound up in the individual contracts between Railtrack and the train operating companies. However, typical penalties can range from between £20 to £100 per minute, so the possible range of consequence could run into millions of pounds, especially for a catastrophic failure. It is, therefore, clear from the risk analysis why Railtrack decided to adopt a strategy to reduce catastrophic failures and the formation of voids rather than one to reduce overtopping.

The decision-making process to adopt a policy to maintain and renovate is best understood through a decision tree developed by

Table C3.8. Results of an analysis of delays during 1995 and 1996 (Hyder Consulting, 1998)

Event	No. of events	Approx. duration of event: min	Approx. delay to trains: min
Catastrophic failure of the seawall	2	63 000	10 500 000
Loss of track support material	1	42 720	72 091
Wave-induced overtopping	15	8000	3000
Landslide material landing on track	1	150	270
Error in safety checks	1	90	100

Thomas and Hall (1992) for considering economic alternatives for existing structures (Figure C3.6).

Although overtopping can cause temporary disruption and there would be an economic benefit to reducing the frequency of overtopping or to improve the accuracy of the warning system, no action has been undertaken to reduce the level of overtopping. It can, therefore, be assumed that the function of the seawall is at least tolerable by Railtrack as long as there is not a catastrophic failure of the seawall. Linking this to the decision tree, Railtrack will accept the consequences of failure due to overtopping, but in terms of the wall's structural strength, the consequences are not acceptable, and to maintain (masonry repairs) and renovate (stepped concrete toe) are currently the best economic alternative. However, in future years the risk of catastrophic failure may reduce to a sufficient level that will place more focus on improving the function of the wall (i.e. reduce delay due to overtopping). The future solution, therefore, may be to upgrade the seawall by either improving the overtopping warning system or trying to reduce the level of overtopping. The overall decision on the level of reduction will be based on economic, engineering and environmental arguments.

In summary, the current levels of risk tolerability accepted by Railtrack are:

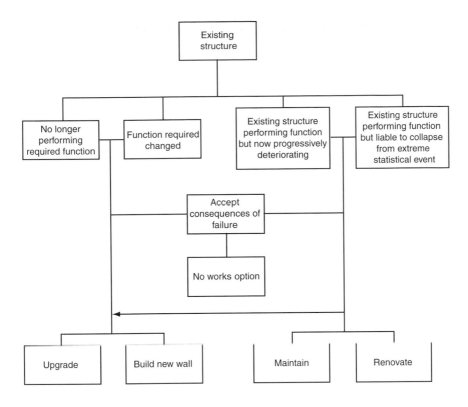

Figure C3.6. Economic alternatives for existing structures

- unacceptable — catastrophic failure or loss of track support material
- tolerable — wave-induced overtopping
- broadly acceptable — minimal penalties.

C3.6. CONCLUSIONS

The main conclusions of the case study are the following.

- The definition of risk involved with a coastal defence set by Railtrack is tailored to its own needs, although it does need to consider other stakeholders. There is little conflict as long as the status quo is maintained or enhancements are made. Risk is expressed as train delay rather than in terms of assets or human life.

- Railtrack has sole responsibility for defining the acceptable level of risk.
- If other stakeholders, such as the local authority had been more active and Government policy had allowed grant funding to a private company, the accepted design may well have been different (e.g. raise the level of the beach along the frontage).
- The selection of potential solutions was developed through a value engineering workshop that involved a large number of people who had different perspectives on the seawall. However, as this workshop was limited to Railtrack personnel only, the views of other stakeholders were not fully represented.
- The detailed design was based on engineering judgement by considering the performance of other structures along the same frontage and locating areas perceived to be at high risk to tackle first, rather than on high level methods such as probabilistic design. Building test sections and monitoring performance were used to check the suitability of the design. The antiquity and uncertainty of the existing construction is the explanation for this type of approach.
- Detailed analysis of the risks allowed design work to focus on the most unacceptable risks and to make a clear decision on what risks were currently tolerable. These decisions may change in the future as risks change with time.

C3.7.   ACKNOWLEDGEMENTS

HR Wallingford would like to thank Railtrack for providing time and information. Special thanks goes to Mr Alan Hardie and Mr Peter Haigh.

C3.8.   REFERENCES

Hyder Consulting (1998). *Railtrack business development — Great Western*. Hyder Consulting, Feasibility study for Dawlish sea defences (unpublished report).

Thomas, R. S. and Hall, B. (1992). *Seawall design*. Butterworth-Heinemann (for CIRIA), London.

Cast study 4.
Grenada sea defences

C4.1. CASE STUDY OBJECTIVES

The objective of this case study is to illustrate the importance of selecting a critical design cross-section or sections such that risk levels remain within the bounds of acceptable (i.e. are not unacceptable nor over acceptable) spatially throughout the design.

C4.2. INTRODUCTION

C4.2.1. Description of the site

The island of Grenada is located in the Caribbean, just north of Venezuela. Like many other islands with a mountainous interior, the coast of Grenada provides a near constant contour that has favoured the siting of the main road and infrastructure links around the edge of the island. However, the coastal road is not low-lying everywhere; in places the coastal road rises some tens of metres above sea level as it traverses steep coastal cliffs. At other sites, the road is cut into more shallow sloping cliff, sometimes close to sea level. In some places, seawalls and revetments have been built in the past to provide protection to the road and cliffs. The length of coastline that covers the area of interest is on the northwest side of the island, near the town of Victoria and stretches for approximately 3 km.

C4.2.2. Description of the problem

The main problem was the ongoing deterioration of the coastal cliff systems, seawalls and revetments to a level that was perceived to be

207

unacceptable, such that the road would soon not be safe to operate. This was caused by waves:

- impacting and undermining existing structures to the point where waves were transmitted under the seawall and up through the road surface
- causing excessive overtopping of low-lying coastal roads so that it was unsafe to travel along
- scouring and undermining the toe of the coastal cliffs and existing seawalls.

C4.2.3. Decision-making process

The Government of Grenada had attempted to provide a solution previously by placing tetrapods and large concrete blocks as emergency measures, which slowed down the problems but did not provide a robust solution. The Government then commissioned DIWI Consult to take a long-term strategic approach to the management of the coastline. As part of that process, DIWI subcontracted Delft Hydraulics to undertake a feasibility study to identify the most appropriate schemes. This involved defining the boundary constraints to the problem, setting the design parameters, generating a number of solutions and selecting the most acceptable solution. A number of schemes were considered from 'do nothing' through to beach replenishment or constructing a new seawall. Posford Duvivier was commissioned in 1995 to take on-board the preferred scheme and to undertake its detailed design and to allocate priorities to the order in which the individual schemes should be constructed. The final design was optimised by using a physical model constructed at Delft Hydraulics before DIWI Consult finally constructed the designed scheme in 1997.

C4.2.4. Description of the solution adopted

The adopted solution comprised various rock armoured revetment structures. The scheme was broken down into three principal types of defence:

- headland defence — design against cliff erosion by wave action
- coast road defence — design against shoreline retreat, destruction of existing defences, overtopping from both operational and catastrophic criteria

- toe defence — design for shoreline retreat, scour at existing defence or cliff face.

C4.3. ACCEPTABLE RISK ISSUES

This section of the case study was undertaken solely by HR Wallingford as part of the research project. Analysis has been used to demonstrate the basic principles set out in the book and should not be considered as accurate design data for other projects.

C4.3.1. Spatial variation of risk

The defences were not continuous along the 3 km frontage, but actually consisted of 20 separate sections/sites. The local wave conditions and other parameters varied from one site to another. This variation could have resulted in a design sizing of rock that would be unique to each of the 20 sites, which would hardly have been a practical approach to the construction of the schemes.

The designer decided to characterise each site by selecting one rock size from a choice of three design rock sizes. This assumption meant that the design rock size was either equal to, or greater than, that derived from theoretical considerations (Van der Meer, 1987) or from physical modelling in a flume. The ratio of the design rock size to that derived was defined as the 'factor of safety'.

The factor of safety was selected in order to:

- ensure that there was a comfortable margin in the design values to cater for tolerance in site parameters and the design methodologies
- avoid overestimation in the selection of rock size.

The quantification of the factor of safety enabled a better informed selection of the design rock size. In practice, size was defined using the nominal diameter (D_{50}) with the design rock sizes defined as 0·85 m, 1·00 m and 1·10 m. The reasons for using this parameter made by the designer were that:

- the rock diameter is (to first approximation) proportional to wave height

- the rock sub-layer grading is a function of D_{50}
- the depth of the rock layers is also proportional to the D_{50}.

Thus, two factors of safety were derived:

- F_1 being the ratio of the design D_{50} to that derived from Van der Meer (1987)
- F_2 being the ratio of the design D_{50} to that derived from physical model testing.

The physical modelling indicated greater stability in the armour than that predicted by Van der Meer and, therefore, resulted in greater factors of safety with respect to the physical modelling for the same design rock size.

In summary, a constant factor of safety along the 3 km frontage was defined by selecting the most appropriate characteristic rock sizes for each site. Typically, the factors of safety were better than $1 \cdot 1$ for F_1 and $1 \cdot 3$ for F_2. The extremes were as shown in Table C4.1.

These results were considered to be satisfactory by the designer. The 'low' case was deemed adequate given that the combined result of F_1 and F_2 was greater than the 10% margin defined. For practical purposes, the 'high' case did not warrant alteration to a smaller rock size.

Overall this example shows that the designer must consider how risk levels may change spatially and either decide to change the design so that the acceptable risk level remains the same or accept the additional redundancy in the structure. The final decision is dictated by a number of forces including cost, time, aesthetics, safety and environmental impact. The designer needs to make his or her best judgement regarding which decision path to select.

Table C4.1.   Factors of safety

	F_1	F_2
Upper limit	1·33	1·57
Lower limit	1·05	1·24

C4.4. ACKNOWLEDGEMENTS

HR Wallingford would like to thank Posford Duvivier for providing the time and data required in this case study. Special thanks must go to Dr Noel Beech, who was the principle contact at Posford Duvivier.

C4.5. REFERENCES

Van der Meer, J. W. (1987) Stability of breakwater armour layers — design formulae. *Coastal Engineering*, **11**, 219–239.

Case study 5.
Tekirdag Port, Turkey

C5.1. CASE STUDY OBJECTIVES

The objective of this case study is to:

- illustrate the concept of acceptable risk relating to the concept of design life and design return periods and the corresponding probability of failure
- illustrate the impact of the selected design event on factors of safety.

These points will be illustrated by focussing the interaction of the selected design wave condition on the stability calculations.

C5.2. BACKGROUND

C5.2.1. Description of the site

An extension to the existing port is being constructed at Tekirdag on the west coast of the Sea of Marmara in Turkey to accommodate Ro-Ro vessels.

C5.2.2. Accepted solution

The project involved construction of new caissons with a cope level of +2·5 m (to datum shown on drawings — see Figure C5.1) and a dredged depth of up to –12 m which will act as a quay for vessels up to 25 000 dwt. Three principal types of caisson were to be

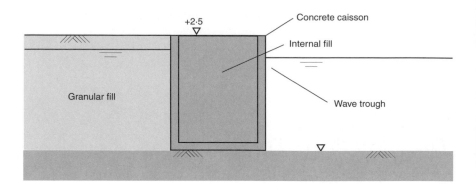

Figure C5.1. Simplified caisson configuration

constructed, with dredged depths of –8 m and –12 m inside the harbour, and –12 m on the external face breakwater. Ground investigations showed that there was a layer of stiff clay situated between –14 m and –17 m, which overlies rock.

Seismic loading at this location was a significant design factor. The north Anatolian Fault beneath the Sea of Marmara is thought to have been accumulating stress since 1766. This built-up stress may trigger a displacement along the fault of over 5 m which is understood to be sufficient to generate an earthquake of magnitude at least as large as that which recently hit northern Turkey.

C5.3. ACCEPTABLE RISK ISSUES

The design process is described above and was influenced by one contractor's preference to construct the quay structure using concrete caissons. However, the generic issues associated with design life and selection of the design loading condition featured strongly in discussions with the client.

C5.3.1. Selection of the design life

Most port structures are designed and constructed for a specific design life. The design life of a structure is taken to be its intended useful life and will depend on the purpose for which it is required.

The choice of the design life is a matter to be decided in relation to each project since changes in circumstances and operational practices can make the structure redundant or in need of substantial reconstruction before the end of its physical life. However, BS 6349 (BSI, 2000) recommends the following values as minimum requirements:

- quay walls — 60 years
- open jetties — 45 years
- superstructure works — 30 years
- dry docks — 45 years
- shore protection works and breakwaters — 60 years
- flood protection works — 100 years.

Based on the above guidance, a 50-year design life was considered satisfactory.

C5.3.2. Definition of design events

The consultant specified the design wave conditions at the commencement of the design process. For the stability calculations, a design wave with a 50-year return period was selected to be the design event as this was considered to be the 'normally' accepted design condition assuming a design life of 50 years.

Overturning and sliding calculations were then performed for eight test design cases, including wave forces and earthquake loading (Table C5.1). The loading categories selected were based on an understanding of the forces that could impact on the structure. Combinations were based on realistic estimations of potential worst case situations that could occur.

For each test condition, the designer made a judgement on the acceptable factor of safety. For example, with Test 4 (bollard and wave forces), the allowable factor of safety was nominally set as 1·2 for both sliding and overturning, based on locally adopted practices (EAU, 1990). The resulting factors of safety for the design return period were 1·51 and 2·02, respectively, for the 50-year return period event. The design, therefore, under the selection criteria adopted was adequate and acceptable. However, if the probability of such an event being exceeded is considered (i.e. the design life equals the return period) then using a 50-year return period seems intuitively optimistic (Table C5.2).

Table C5.1. Design cases used in stability calculations of the caisson

Loading test no.	Description of design case	Dead load	Live crane, surcharge	Bollard force	Negative wave force	Earth fill	Earthquake loads	Difference in water level of 1 m
1	Normal	+	+	–	–	+	–	–
2	Bollard	+	+	+	–	+	–	–
3	Bollard and difference in water level	+	+	+	–	+	–	+
4	Bollard and negative wave force on caisson	+	+	+	+	+	–	–
5	Earthquake 1	+	+	+	–	+	+	–
6	Earthquake 2	+	+	–	–	+	+	–
7	Earthquake 3	+	+	+	–	+	+	–
8	Earthquake 4	+	+	–	–	+	+	–

Table C5.2.   Probability of exceedance of a design return period during a design life

Return period: years	Design life: years						
	2	10	**50***	100	200	500	1000
2	0·63	0·99	**1·00**	1·00	1·00	1·00	1·00
10	0·18	0·63	**0·89**	1·00	1·00	1·00	1·00
50*	**0·04**	**0·18**	**0·63**	**0·86**	**0·98**	**1·00**	**1·00**
100	0·02	0·10	**0·39**	0·63	0·86	0·99	1·00
200	0·01	0·05	**0·22**	0·39	0·63	0·92	0·99
500	0·00	0·02	**0·10**	0·18	0·33	0·63	0·86
1000	0·00	0·01	**0·05**	0·10	0·18	0·39	0·63

* Design life and design return period assumed by the consultant for Tekirdag Port.

$P = 1 - \left(1 - 1/TR\right)^{n}$ where TR is the design return period and n is the design life.

Structures designed to withstand large events tend to work out to be more expensive than weaker structures for which the cost of periodical repair has been included. By optimising whole-life costs against benefits, it may be possible to establish an acceptable level of risk. This type of cost optimisation is, however, difficult to achieve in quay structures, as the degree of damage is difficult to predict. It was suggested by the reviewer of the design that these type of structures are typically often designed with a design life of approximately 50 years and a corresponding minimum design return period of between 100 and 200 years (HR Wallingford, 2000). The probability of exceedance would therefore be reduced to 39% and 22%, respectively. It was noted that it was the port's responsibility to define the level of risk that they wish to accept; if, however, they wish to accept more internationally based working practices, then the reviewers' recommendation of a 1:200-year return period should be selected.

With this in mind, the designer made some checks that considered the sensitivity of the factor of safety to a change in the design return period used for Test 4. In this particular example, water level remains reasonably static and the change in wave height between the 50- and

100-year return period is reasonably small (from 3·1 m to 3·3 m) such that there is little impact on the factor of safety for both overturning and sliding (Table C5.3).

So, in this particular design, the assumption of using a 50-year return period rather than a 100-year or a 200-year return period was not critical. In fact, the earthquake loading conditions were more critical than the hydraulic elements of the design.

This illustrates the point that the designer must clearly understand the potential variation in loading conditions and how they may impact on the designs safety level. In the Sea of Marmara in Turkey, fetches are limited and tidal variations are small, but if, however, the same design was to be undertaken on the south-west coast of the UK, then the influence of wave height and water level would clearly be more significant because of the greater tidal range, including storms, surges, and larger waves from the Atlantic. This is illustrated by playing with the input parameters of the stability calculations for overturning (Table C5.4).

The analysis highlights two interesting points (Figure C5.2):

- a significant change in wave height had little impact on the overall factor of safety with convergence as the wave height reduces
- a change in the depth of water had the most impact on the factor of safety. If water level overtime was significantly variable then a joint probability study of waves and water levels may be a useful exercise to select the appropriate design event. Care needs to be taken when considering multiple design parameters that the correct tool or technique is used to analyse the problem (see Case study 6).

Table C5.3. Impact on the factor of safety if another design return period selected

Return period: years	Factor of safety	
	Sliding	Overturning
10	1·66	2·06
25	1·65	2·05
50	1·64	2·04
100	1·62	2·03

Table C5.4. Design parameter changes to overturning stability analysis

	Return period: years			
	10	25	50	100
Original wave heights: m	2·5	2·9	3·1	3·3
Original water depths at structure: m	12·5	12·5	12·5	12·5
Change in wave height, depth static: m	2·3	No change	3·4	3·6
Change in water depth, original wave heights: m	12·0	No change	13·0	13·5

     This analysis illustrates that when selecting the design return period the designer needs to have regard to the potential sensitivities to change that may occur to the design parameters. Other sensitivities that may need to be checked, depending on the particular project, are

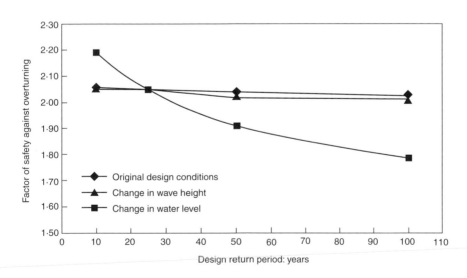

Figure C5.2. Impact on factor of safety due to variations in wave height and water level

issues such as climate change and reduction in beach levels, for example. However, in this particular case the selection of a specific design return period had little impact on the overall scheme of things. It is essential, therefore, that the client has the opportunity to understand clearly the basis of the design and the levels of risk that they are being asked to take on. In this case, the consultant was able to do this and agree the most appropriate design approach.

C5.4. ACKNOWLEDGEMENTS

HR Wallingford would like to thank contractor Akport and designer Dolfen for agreeing to provide information for this case study. The Akport representatives were Fuat Ozbekli, General Manager, and Cenk Mehmet Anil, Marketing Manager at Akport. The Dolfen representatives were Bulent Bilgili and Yasemin Ozgen. The HR Wallingford Project Manager was Ian Cruickshank.

C5.5. REFERENCES

British Standards Institution (BSI) (2000). *Code of practice for maintaing structures — Part 1. General criteria.* BSI, London. BS 6349-1.

EAU (1990). *Recommendations of the Committee for Waterfront Structures, Harbours and Waterways, Society for Harbour Engineering and the German Society for Soil Mechanics and Foundation Engineering*, sixth edition. Ernst & Soln, Berlin.

HR Wallingford (2000). *Tekirdag Port, Review of structural calculations.* HR Wallingford, Wallingford, HR Report EX 4271.

Case study 6.
West Bay coastal defence and harbour improvement scheme

C6.1. CASE STUDY OBJECTIVES

The objective of this case study is to demonstrate how different analysis techniques of coastal wave-induced overtopping can have a major impact on the assessment of flood risk and the resulting design

C6.2. BACKGROUND

West Bay is a small port located on the West Dorset Coast and is at the western end of the Chesil Bank (Figures C6.1 and C6.2). The present day harbour was constructed in 1740 and is accessed by a 110 m long by 12 m wide navigation channel formed by two aligned piers. West Bay is also flanked by two different coast defence systems, to the east of the harbour lies East Beach a large shingle barrier and to the west lies West Beach with a masonry seawall and a beach that has been lowering by approximately 4 mm per year.

HR Wallingford was commissioned by West Dorset District Council to undertake a strategy study for the frontage at West Bay that considered both improving navigation into the harbour and reducing the risk of flooding to an acceptable level.

The overall strategy study and design work considered a wide range of issues from the demolition of a Grade II listed structure and the environmental impact of the preferred solution. For the purposes of the case study, we have concentrated only on the overtopping discharges of one section of the seawall along West Beach. A sketch

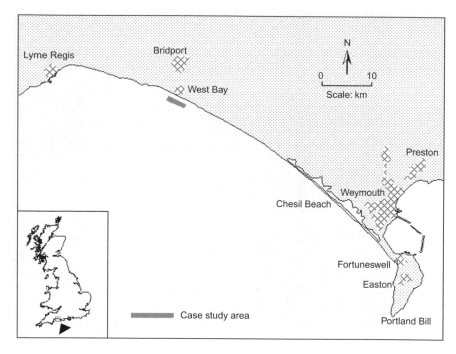

Figure C6.1.   Geographical location of West Bay

of the seawall is presented in Figure C6.3. The preferred solution for this section was to increase the beach levels in front of the wall to +4·5 m ODN.

C6.3.   ASSESSMENT OF OVERTOPPING RATES AT WEST BAY

C6.3.1. Different analysis methods used

Engineers often calculate the overtopping discharge for a particular return period in order to assess the potential risk of flooding or structural damage. Traditionally, this has usually been calculated for the worst storm event for a required return period, which is selected by testing a number of different sea conditions (wave height, wave period, water level combinations) with the same joint return period.

Figure C6.2.   Aerial view of West Bay

This method is simple to apply and analyse, but, by selecting only a specific point, makes the clear assumption that the return period of the loading event equals the return period of the response event. However, in the majority of cases, parameters such as the geometry of the wall, the breaking point of the wave, etc., can produce a higher overtopping discharge for a more frequent sea condition.

For risk assessment purposes, it is, therefore, better to work with a larger dataset of loading conditions. Techniques now exist to produce a synthetic dataset that has frequency distributions equivalent to those of the loading sea conditions being considered. Each record in the synthetic dataset can then be used to calculate a corresponding overtopping rate. A frequency analysis of the overtopping discharges can then be carried out to generate the univariate distribution and thus find out the return period or frequency of each discharge. The great advantage of this data-rich approach is that it enables multivariate

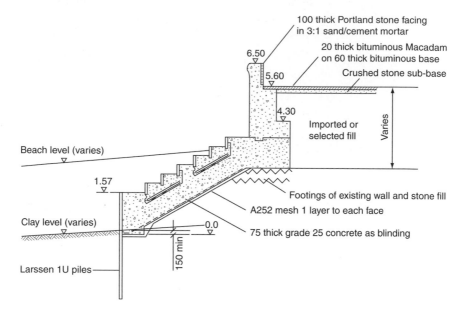

Wall construction
Constructed: 1968–1969
Contractor: Ashworth Construction Ltd
Length: 50 m

Strengthening
Constructed: 1981–1983
Contractor: Amey Roadstone Ltd
Consultant: Dobbie & Partners Ltd

100 thick Portland stone facing
in 3:1 sand/cement mortar

20 thick bituminous Macadam
on 60 thick bituminous base

Crushed stone sub-base

6.50

5.60

4.30

Imported or
selected fill

Varies

Beach level (varies)

1.57

Footings of existing wall and stone fill

A252 mesh 1 layer to each face

Clay level (varies)

0.0

75 thick grade 25 concrete as blinding

150 min

Larssen 1U piles

Figure C6.3.   Example cross-section of the seawall analysed for overtopping

loading condition distributions to deliver a single univariate distribution of response (in this case overtopping discharge), enabling clearer risk decisions to be made.

At West Bay, this analysis was done by producing a dataset that represented 10 000 years of data with a record for every high tide in a year with a corresponding wave height (i.e. 707 tides × 10 000 years = 7 070 000 events).

C6.3.2. Results

The discharge rates were predicted for both techniques using the same formulae. The predictions using the two different methods are

Table C6.1.   Summary of overtopping discharge predictions

Return period: years	Loading event: l/s/m	Response event: l/s/m	Increase multiplier
0·1	0·02	0·07	3·50
1	0·3	0·7	2·33
2·5	0·8	1·4	1·75
10	1·7	3·0	1·76
100	5·4	8·5	1·57
2000	16·1	26·4	1·64

Table C6.2.   Difference in likely return period of response if loading design event used

Return period in terms of loading: years	Resulting return period in terms of response: years	Reduction multiplier
1	>0·1	—
2·5	1	0·4
10	2·5	0·25
100	20	0·2
2000	300	0·15

summarised in Tables C6.1 and C6.2. If reliance on the loading condition as a design event is assumed, then the actual true response with the same return period may be out by a factor of between 1·5 and 3·5 or even higher for more frequent events. Similarly, if the response from the loading event was assumed, then the resulting return period in terms of response would be significantly lower (i.e. more risky for the client).

C6.3.3. Potential impact of different analysis techniques

The main aim of the study was to assess flood benefits in order to justify capital works to protect West Bay. In this particular situation, even though West Bay had a history of flooding, the benefits or assets at risk were limited, so a more robust risk assessment approach was required in order to maximise benefits.

A larger dataset was useful when considering more detailed design issues such as sensitivity of sea level rise or further lowering of beach levels. Was this usefulness economically viable given the extra resource required to undertake the work? In the long term, the assessment was viable because if the traditional method had been adopted there would be more uncertainty regarding the structural response of the wall, which would have to be accommodated somewhere in the design. A partial factor of safety could have been applied or a higher return period selected for design purposes, but without detailed risk assessment it would have been very difficult to make an accurate estimate of such a safety factor. Also, the benefits assessed would not have been as high and, therefore, resulting in the potential for a cheaper but less acceptable solution.

In this particular case, a physical model was also used to optimise beach crest levels in front of the wall in order to reduce overtopping to an acceptable level. Using the detailed method allowed more confidence in the selection of design sea conditions used in the physical model and the observations made were then used to improve the certainty of the risk assessment method without any further analysis.

C6.4. CONCLUSIONS

The main conclusions are as follows.

- A risk assessment approach was developed during the study that encompassed the statistical wave and water level climate at West Bay. As a result, the accuracy of predicting the risk of a certain response was improved by taking a more statistically rigorous approach.
- This approach was useful in this situation because the return period of the loading event was not directly linked to the return period of the response event. The designer needs to take

professional care to assess the risks appropriately and to apply a certain level of engineering judgement on the results, depending on the level of uncertainty.

- If a traditional approach had been utilised then the level of risk communicated to the client and other stakeholders would have been higher, with the client accepting more risk than expected.

C6.5. ACKNOWLEDGEMENTS

HR Wallingford would like to thank West Dorset District Council for agreeing to provide information for this case study. The West Dorset District Council representatives were Mr Keith Cole, Engineering Manager, and Mr David Joy, Project Manager. The HR Wallingford Project Manager was Paul Sayers.

Case study 7.
Design of scour protection for bridges

C7.1. OBJECTIVE OF THE CASE STUDY

The objective of this case study is to demonstrate the potential for a more robust assessment of acceptable risk levels on a project by project basis.

C7.2. SETTING A DESIGN EVENT FOR SCOUR PROTECTION

There is considerable literature on the study of bridge structural safety and the problems involved in applying it. Analysis (by Smith, 1976) showed that a large number of historical bridge failures from 1877 to 1975 were due to flooding (Table C7.1). This percentage increases dramatically when only bridges built over watercourses are considered. Among 86 bridge failures analysed from 1961, the percentage due to flooding rose to 55%. Scour is, therefore, one of the most common failure mechanisms. Note that Smith did not base his analysis on any formal statistical sampling but it provides a useful indication of the importance of scour in bridge design.

The majority of bridges designed in the UK come under the auspices of the Highways Agency. The standard design procedures set out by the Highways Agency state that the design life of a bridge should be 120 years and that scour protection be designed to withstand a 200-year flood (Highways Agency, 1994). Guidance by the Federal Highways Institute in the USA states that a 100-year flood is sufficient for bridge design.

On first inspection, a 200-year flood seems reasonable as the indicative standards for flood protection in the UK range between

Table C7.1.   Summary of bridge failures from 1877–1975

Cause of failure	No. of failures	Total failures: %
Inadequate or unsuitable temporary works or erection procedure	12	8
Inadequate design in permanent material	5	3
Unsuitable or defective permanent material or workmanship	22	15
Wind	4	3
Earthquake	11	8
Flood	70	49
Fatigue	4	3
Corrosion	1	1
Overload or accident	14	10
Total number of failures	143	100

100 and 200 years. However, these are accepted over a much smaller design life of approximately 60 years and, as a result, such a flood has a much smaller probability of being exceeded during the design life. If the design life of the bridge is set at 120 years then for a 100-year flood the exceedance probability is 0·7 and for a 200-year flood this only reduces to 0·45. At present, the selection of the hydraulic design condition is both arbitrary and at a risk level such that scour will inevitably be the most frequent cause of failure.

If one assumes that the design conditions for scour are exceeded, then the scour protection will fail causing the bridge to collapse. The replacement of the bridge and the associated disruption costs of the bridge collapsing, plus the cost of a higher standard of scour protection, can be easily estimated. In general, the expected whole-life cost risk of the bridge will decrease as the design conditions are based on higher return periods (i.e. become less likely). Risk can be defined as:

$$Risk = P_e(Tr) \times NPV$$

where $P_e(Tr)$ is the probability of a particular design return period occurring during the design life of the bridge and NPV is the net present value of a failure (i.e. the summation of original cost, replacement cost, disruption costs, etc.).

In order to resist higher return periods, the cost of scour protection will have to increase. At some point these two curves will cross. The intersection of these two curves will reflect the optimum design return period for scour protection works. A hypothetical example has been used to illustrate this point and is described below.

The cost of a typical bridge over a watercourse with a single span of 18 m is within the region of £800 000 and £1 200 000. Disruption costs if a bridge collapses could be enormous depending on the nature of the crossing, but a crude assessment of £500 000 seems reasonable. The cost of scour protection can range from £50 000 to £500 000 or more depending on the extreme flow velocities in the channel. For the purposes of this example, the following assumptions have been made:

- the original capital cost of the bridge is £800 000
- the capital cost to rebuild the structure is £1 000 000
- the likely disruption costs are £500 000
- the cost of scour protection ranges from £50 000 for a 1-year return period and £500 000 for a 10 000-year return period.

The risk profile for three different design lives show that the optimum scour design return period should vary significantly. The 200-year event is acceptable if the design life is only 25 years but an event up to the 800-year return period should be used if the full design life of 120 years is accepted (Figure C7.1). As part of the design process, it is important to understand the impact of design life and design event; if ignored, the risk level accepted may actually fall short of the performance required.

This simple analysis shows that the economic consequences of a failure dictate the level of safety that should be accepted. However, care needs to be taken by the designer to ensure that the stregthening of one parameter is not achieved at the negative consequence of another. For example, as the bridge is protected against higher return period storms, this investment may be wasted because water levels rise so high that the deck becomes buoyant and fails before the scour design return period occurs. The designer needs to have a clear understanding, therefore, of the potential failure modes and their

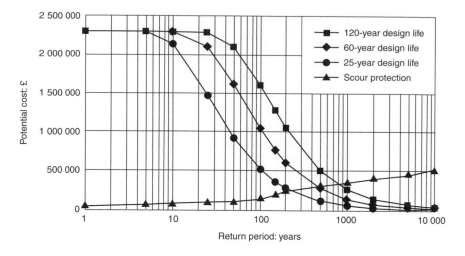

Figure C7.1. Risk of failure against the capital cost of scour protection

interaction. For each project, the risk level required will be different and, as a consequence, should be quantified for that particular situation using published design guidelines only as an indicative starting point.

C7.3. ACKNOWLEDGEMENTS

HR Wallingford would like to thank Dr Roger Bettess of HR Wallingford for providing the initial concept of the case study and Mr Alan Hardie of Railtrack for providing indicative costs of simple bridge construction and scour protection.

C7.4. REFERENCES

Smith, D. W. (1976). Bridge failures. *Proceedings of the Institution of Civil Engineers*, **60**, Part 1, August, 367–382.

Highways Agency (1994). *Design manual for roads bridges*, Vol. 1, Section 3, Part 6, BA59/94 — The design of highway bridges for hydraulic action. HMSO, London.

Case study 8.
Acceptance of higher risk levels by the client

C8.1. OBJECTIVE OF THE CASE STUDY

The objective of the case study is to demonstrate an example where a client, in this case the National Rivers Authority (NRA), has decided to accept a higher than normal level of risk in order that the project could be completed and the construction cost minimised.

C8.2. EXAMPLE OF A PROJECT

The project included improvements to an existing embankment on a tidal river. The existing embankment comprised mainly silty sand that had been shaped over the years to 1 in 2 slopes and a 4 m wide crest. Parts of this embankment were eroded and suffering distress, including reduced top height. The initial design for the works followed basic stability calculations assuming uniform ground beneath the embankment and assuming uniform material in the embankment. This would restore the height and condition of the embankment in the project to the same profile as the adjacent and recently improved embankments.

During construction, which included building up layers of geotextile and imported sand, the foundation of the embankment considerably settled and beyond that which had been anticipated. Effectively, the foundation had failed under the weight of the construction activity.

The design consultant for the project reviewed the position, undertook further soil analysis and calculated a revised embankment profile in order to accommodate the changed ground conditions and to retain the same top height and defence standard for the embankment. A target factor of safety of 1·5 was used against further slip circle failure.

The resulting size of the new design more than doubled the footprint for the embankment foundation and, therefore, doubled the volume of fill required in order to achieve the same factor of safety. The consequences of such a design were not only to increase the quantities, and, therefore the cost, but also to increase considerably the amount of land required for the embankment and would also have involved dealing with several properties which would otherwise be engulfed by the fill.

The NRA could not accept these consequences of the redesigned embankment and discussed opportunities for a more modest design with the consulting engineer. One of the sticking points appeared to be that the fill in the existing embankment was of a variable nature and, therefore, did not readily conform to soils analysis and inclusion within normal computerised design programs. Back analysis of the adjacent and satisfactory embankment was suggested in order to obtain apparent soil parameters for redesign of the embankment in the project. This was only marginally successful.

The most significant difference between the computer analysis of the existing embankment and that of the proposed redesigned embankment was that the factor of safety for the existing adjacent embankment appeared to be little above unity, whereas the redesigned embankment with the larger footprint could only achieve a factor of safety of 1·35 compared to the desired 1·5.

On the basis that the adjacent existing embankment was standing satisfactorily and had stood the test of several years since its construction, it seemed reasonable that if the project embankment could be constructed to the same profile with a factor of safety of approximately unity, then, once constructed, it had a more than reasonable chance of survival.

The consulting engineer was not happy with this suggestion; his professional insurance required him to follow codes and recommended factors of safety (i.e. 1·5 in this case) but clearly this did not align with the client's objectives. In order to progress, NRA wrote to the consulting engineer and, for this specific project, instructed that a factor of safety of just greater than unity would be satisfactory and that there would be no recourse to the consultants' professional indemnity insurance should a further failure occur. All the consultant was now left to do was to design an embankment which met the height and shape criteria required within the footprint available and achieve a factor of safety of greater than one. This was done and, in fact, the factor of safety was calculated to be 1·3. The embankment was constructed in 1995 and stands today in good condition without any distress.

In conclusion, while it may only be within the gift of the client to accept a reduced standard of construction compared to that which might normally be provided by a consulting engineer, there is clearly scope for adjustment in the balance of risk as a consequence of the design. It was also surprising that when relieved of the need to achieve a factor of safety of 1·5, it was found that a factor of 1·3 was in fact readily available. Consultants have a duty to their clients to provide them with a robust design, however, they also have duties to protect the client's overall interests and to recommend the possible use of a reduced factor of safety in cases where there are good reasons to do so.

C8.3. ACKNOWLEDGEMENTS

HR Wallingford would like to thank Mr Gordon Heald of the Environment Agency for providing the information for this case study.

Index